High Performance Clocks with Special Emphasis on Geodesy and Geophysics and Applications to Other Bodies of the Solar System

Rafael Rodrigo • Véronique Dehant •
Leonid Gurvits • Michael Kramer • Ryan Park •
Peter Wolf • John Zarnecki
Editors

High Performance Clocks with Special Emphasis on Geodesy and Geophysics and Applications to Other Bodies of the Solar System

Previously published in *Space Science Reviews* in the Topical Collection "High Performance Clocks with Special Emphasis on Geodesy and Geophysics and Applications to Other Bodies of the Solar System"

Springer

Editors
Rafael Rodrigo
International Space Science Institute
Bern, Switzerland

Véronique Dehant
Royal Observatory of Belgium
Brussels, Belgium

Leonid Gurvits
Joint Institute for VLBI in Europe (JIVE)
Dwingeloo, Drenthe, the Netherlands

Michael Kramer
Max-Planck-Inst für Radioastronomie
Bonn, Nordrhein-Westfalen, Germany

Ryan Park
Jet Propulsion Laboratory
Pasadena, CA, USA

Peter Wolf
Paris Observatory
CNRS
Paris, France

John Zarnecki
International Space Science Institute
Bern, Switzerland

ISSN 1385-7525 Space Sciences Series of ISSI
ISBN 978-94-024-1656-5 ISBN 978-94-024-1566-7 (eBook)
https://doi.org/10.1007/978-94-024-1566-7

Cover Image: CNES/ ill. David Ducros

Printed on acid-free paper

This Springer imprint is published by the registered company Springer Nature B.V.
The registered company address is: Van Godewijckstraat 30, 3311 GX Dordrecht, The Netherlands

Foreword

The ESA High-Level Science Policy Advisory Committee (HISPAC) was established in 2007 by the then ESA Director General, Jean-Jacques Dordain. Its purpose is to serve as the highest level of the ESA Science Advisory Structure, advising the DG directly on long-term science policy for all programmes of the Agency. Its remit is *'To reflect upon interdisciplinary science and technology themes and ideas "unconstrained by considerations of funding, ESA's internal Directorate structure, the existing approved programme and politics'*. Its vision is to promote a *'One-ESA'* and *'One Science Programme'*, an Agency-wide coherent and comprehensive approach to science and technology as a whole.

In 2012, HISPAC completed the definition of a set of *'Grand Science Themes'*. These provide an overarching view of science topics for the next 30 to 50 years that should be addressed by future missions of the Agency. The objective was to help focus ESA's key science and technology priorities. The guiding principles were that the themes should be scientifically rich, inspirational and extensive. They are synergistic across the Agency's different scientific programmes, and cross-linked with enabling technologies of relevance and value across ESA. The four themes are: *'Terrestrial and Cosmic Climate'*; *'Understanding Gravity'*; *'Life in the Universe'*; and *'Cosmic Radiation and Magnetism'*.

The *Grand Science Themes* provide on the one hand a synthetic view, emphasizing the cross-disciplinary aspects and thus providing a good instance of the *'Science-led one-ESA'* approach advocated by the DG. On the other hand, at a lower level they offer the possibility of substantive detail on specific scientific disciplines, thus allowing for practical planning of required technology developments, as well as bridging current and future planning of the Agency's programmes.

It is at this lower level that HISPAC has sought external advice. Because of their potential implications on future programmes and budgets, the scientific underpinning of the *Grand Science Themes* can benefit greatly from independent inputs from the expert community at large. A key constraint is to avoid interfering with the operational process by which the Agency's programmes are defined and selected. The objective is to 'pump prime' new ideas and possibilities, and make them available so that the science communities of the ESA member states can draw upon them as they wish.

Various options were considered by HISPAC for the means to achieve this. The approach chosen was to draw upon the expertise and support of the International Space Science Institute (ISSI) in Bern. ISSI was established in 1995 with the purpose of allowing scientists from all over the world to meet in a multi- and interdisciplinary setting to reach out for new scientific horizons. Its tried-and-tested methodology for organizing Workshops and Fora, and for publishing books to provide an authoritative account of the state of play, was readily adaptable to suit HISPAC's purpose.

In 2013 a pilot phase was carried out, consisting of a scientific Forum on *'Understanding Gravity'*. This took place at ISSI in Bern on the 3rd and 4th of December. A total of 24 participants, including scientists and technologists, reviewed the possibilities for cross-fertilisation between the programmes in different ESA directorates. The group endorsed HISPAC's view that there are large benefits to be gained by detailed transfer of technical information between the different areas of science within ESA. They concluded that opportunities exist for cross fertilisation between science areas as diverse as fundamental physics and Earth Observation. They recommended that HISPAC follow up with a more extensive Workshop and suggested possible topics.

At its meeting in Frascati in October 2014, HISPAC decided to entrust ISSI with the organization of a follow-up Workshop entitled *High Performance Clocks, with special emphasis on Geodesy and Geophysics and applications to other bodies of the Solar System.* The concept was to bring together scientists from space research and ground development, from time transfer techniques and radioscience missions, from the time-frequency community, and from both the geodesy-geophysics and solar system exploration communities.

Based on the outcome of the earlier forum, HISPAC identified the following topics to be covered:

- Clocks, Time Transfer, Technology
- Applications of high-precision clocks in Earth Sciences
- Applications of high precision clocks in Fundamental Physics
- Applications of high-precision clocks in Solar System exploration

The 'deliverables' were specified as a clear identification of the most promising future scientific research areas and potential ESA missions in the different disciplines covered by the workshop, and areas of possible synergy in the science and technology domains between the different scientific topics or missions proposed.

The Workshop took place between November 30 and December 4, 2015. Attended by 50 participants from a broad range of disciplines and backgrounds, it was an intensive and stimulating experience. The record documented in this volume provides a detailed account of the deliberations and conclusions, and hence delivers a sound basis for future developments within the '*Understanding Gravity*' theme. On behalf of HISPAC and ESA, I wish to express my thanks to the ISSI Director and staff for the highly successful manner in which the Workshop was planned and run.

Prof. Chris Rapley CBE

Professor of Climate Science
University College London
Chair of HISPAC

Contents

Preface

Véronique Dehant[1] · **Leonid I. Gurvits**[2,3] ·
Michael Kramer[4,5] · **Ryan Park**[6] · **Peter Wolf**[7] ·
John Zarnecki[8,9] · **Rafael Rodrigo**[9]

1 Role of Clocks in Fundamental Physics

As described by one of the papers in this volume (Delva et al. 2018), there are two fundamental theories of physics: the Standard Model of particle physics that describes electromagnetism and the (strong and weak) nuclear interactions, and General Relativity that accounts for all gravitational phenomena. The dream of all physicists is to unify these two theories, describing as much as possible of the observed Universe and solving the still-remaining open issues of both theories and observations (e.g. the nature of dark matter and dark energy). Most physicists agree that there is still one theory that remains to be discovered in that respect. Attempts at such a unified theory lead to tiny violations of the basic principles of these theories, in particular of the Einstein Equivalence Principle (i.e., equivalence of gravitational and inertial masses). Delva et al. (2018) review a number of alternative theories and present their pros and cons, as well as the possibilities that can be offered by further

✉ V. Dehant
Veronique.Dehant@oma.be

1 Royal Observatory of Belgium, Brussels, Belgium

2 Joint Institute for VLBI, ERIC (European Research Infrastructure Consortium), Dwingeloo, Drenthe, the Netherlands

3 Department of Astrodynamics and Space Missions, Delft University of Technology, Delft, the Netherlands

4 MPI für Radioastronomy, Bonn, Germany

5 Jodrell Bank Centre for Astrophysics, University of Manchester, Manchester, Germany

6 Jet Propulsion Laboratory, Pasadena, CA, USA

7 SYRTE, Observatoire de Paris, PSL Research University, Sorbonne Universités, UPMC, Univ. Paris 06, LNE, CNRS, 75014, Paris, France

8 Open University, London, UK

9 ISSI, Bern, Switzerland

more-precise observations, in particular by using high performance clocks. The search for discoveries at the experimental level requires high accuracy measurements that need the use of the outstanding performance offered by a time/frequency metrology. The role of clocks in these observations is fundamental and their use outside the Earth's gravitational field is also fundamental. Clocks on Earth are limited by the knowledge of the local Earth potential and of its fluctuations. As a matter of fact, the SI units themselves rely on a time/frequency realization.

Precise clocks can be used not only to verify the Einstein Equivalence Principle via precise redshift measurements or tests of gravity at Solar System scale. On Earth, the best reference standard clocks are at the 10^{-18} precision level and, in space, they are at 10^{-16}. Such high-performance clocks can be compared over large distances and in varying gravitational fields when used in space, which makes them very valuable for fundamental physics experiments. Such experiments are described in this topical collection by Delva et al. (2018). In particular, ultra-stable microwave or optical clocks in space from the ESA ACES mission are (1) PHARAO, a laser cooled Cesium clock, and (2) SHM, a space hydrogen maser. They are related to clocks on Earth with an improved version of the Microwave link (MWL) as well as an optical link for high performance space-ground clock comparisons. The launch of the ACES/PHARAO platform aboard a Dragon capsule of SpaceX (CRS-13) is scheduled for mid-2018 and its use is foreseen onboard the International Space Station (ISS) in the 2020–2025 timeframe. Improvement is expected to come from the accuracy of the clock (at the 10^{-17} level in fractional frequency) and the better performance of the time/frequency transfer links. Many new experiments related to fundamental physics in space are described by Delva et al. (2018). This is a very active field that relies and thrives on technological developments, in particular of clocks, long distance radio and optical links, and inertial sensors (e.g., accelerometers and gyroscopes).

2 Role of Clocks in Fundamental Geodesy

Atomic clocks are also outstanding tools for experiments and observations performed in fundamental geodesy such as VLBI (Very Long Baseline Interferometry), pulsar timing, navigation, etc. Clocks are currently used for this purpose on Earth. However, precise measurements from space with a relay satellite are foreseen where clocks will play a fundamental role.

One of the missions that is expected in the future is the E-GRASP/Eratosthenes mission, a present proposal for the ESA's Earth Explorer Mission EE-9. E-GRASP/Eratosthenes is dedicated to improving and homogenizing time and space references on Earth and, more directly, to realizing the terrestrial reference system with an accuracy of 1 mm and a long-term stability of 0.1 mm/year, which have so far not yet been achieved. The E-GRASP/Eratosthenes platform assembles all the geodetic techniques used for the realizations of reference frames and timescales, themselves used in many high-accuracy geophysical applications. For instance, quantifying sea-level change or the effects of ice melting using altimetry strongly depends on an accurate determination of the position of continental or island reference stations, such as those located at tide gauges, as well as the ground stations of the tracking networks. Previous terrestrial reference systems have been realized by a combination of the four space geodetic techniques (VLBI, GNSS (Global Navigation Satellite Systems), DORIS (Doppler Orbitography and Radio-positioning Integrated by Satellite), and SLR/LLR (Satellite and Lunar Laser Ranging)) to optimally use the strengths of each technique. The Global Geodetic Observing System (GGOS) was launched by the International Association of Geodesy (IAG) with the goal of providing a consistent high-quality reference frame for observing changes in the Earth system. The documents delivered by IAG demonstrate the need for the above-mentioned accuracies. This requirement is especially driven by the fact that stability in the present global terrestrial reference frame (TRF) of about 0.5 mm/year is one of the major uncertainties that is mapped in the determination of global sea level rise. The importance of a highly accurate, stable and sustainable reference frame has even been recognized by the A/RES/69/266 United Nations Resolution.[1] The full and rigorous combination of all space geodetic techniques rely on including the common clock of high precision in space. The mission is undergoing a down-selection at present for the future EE9 Explorer Mission. Even if the mission is not selected, the scientific goals remain of high importance for the scientific community as well as for the society.

It is worth mentioning that high-precision clocks in the context of the realization of reference frames and time scales used in geodesy as mentioned above, will also soon require the re-definition of time scales or, at least, their clarification to ensure their continuity and consistent use in practice, as discussed by another paper in this volume (Müller et al. 2018).

Let us come back to the highly precise time and stable reference frequencies in the frame of space geodesy techniques themselves. It is a fundamental requirement that VLBI ground stations or the GNSS and DORIS satellites are equipped with high-precision clocks. The precision of these clocks evolves with time, and their number used in the satellites increases as well. In particular, the new generation of GPS satellites are equipped with cesium and rubidium clocks/atomic frequency standards. The Galileo satellites carry four atomic clocks each to ensure strong, quadruple redundancy of the timing subsystem: two Rubidium Atomic Frequency Standard (RAFS) clocks and two Passive Hydrogen Maser (PHM) clocks. It has recently been reported that anomalies have been noted in the atomic clocks serving Europe's Galileo satellites. A few RAFS & PHM clocks unexpectedly failed showing consistent signatures, linked to probable short circuits or long turn-off periods, which is presently under investigation. The only implemented to date example of an Active Hydrogen Maser aboard a scientific satellite mission is the Space VLBI mission RadioAstron/Spektr-R, which has demonstrated promising preliminary results in the gravitational redshift experiment aimed at verification of the Einstein Equivalence Principle (see Litvinov et al. 2017, Phys. Letters A, in press, arXiv 1710.10074).

Satellite Laser Ranging (SLR) is another geodesic technique, which differs from all other applications like VLBI, GNSS DORIS by the fact that it is a two-way transfer technique. That means that there is no need for a clock synchronization process between both ends of the distance covered by the measurement technique. The paper by Schreiber and Kodet (2018) discusses the role of time in SLR and time transfer. In particular, laser optical techniques for optical clocks over distances of thousands of km and for time transfer are discussed and shown to be of high importance. Also very promising, although not discussed in the paper, is free-space optical telecommunication. On Earth, this optical communication technology uses light propagating in free space to wirelessly transmit data for telecommunications or computer networking. It is presently being studied also for interplanetary links. We believe that new emerging technologies, and in particular those associated with optical links such as Interplanetary Laser Ranging (ILR)—addressed by the present paper of Dehant et al. (2018), active and passive retro-reflectors, Light Detection And Ranging (LIDAR) can overcome the limitations of radio technology and improve the measurements' precision significantly.

[1] http://www.un.org/en/ga/search/view_doc.asp?symbol=A/RES/69/266.

Very Long Baseline Interferometry is a radio astronomy technique developed for observations of celestial radio sources. It is used to study these sources as well as realize the celestial reference frames. Currently, hydrogen maser clocks are used as a standard ingredient of VLBI instrumentation enabling coherent processing of radio signal of celestial sources received at widely separated radio telescopes. Improving the quality of VLBI observations especially at mm- and sub-mm-wavelengths would require getting frequency stability of local oscillators beyond achievable presently by Hydrogen maser frequency standards. In this volume, Nothnagel et al. (2018) also discuss why this is, and identified four areas where good clock performance is needed are considered: coherence, geodetic parameter estimation, correlator synchronization, and UT1 determination. Both traditional VLBI where extra-galactic radio sources are observed, as well as observation of man-made artificial radio sources emitted by satellites or spacecraft are considered.

3 Role of Clocks in Gravity Field Determination

As explained above and detailed in this volume by Müller et al. (2018), time measured by an ideal clock, when compared to another distant clock, crucially depends on the gravitational potential and velocity of the clock according to general relativity. As clock precision is now at the level of 10^{-18} in fractional frequency, they are sensitive to very small differences of the gravitational field, which makes them very useful for understanding the relativistic gravitational physics. According to general relativity, the relative frequency shift between two clocks is directly related to the difference in the values of the gravity potential at the positions of the clocks. This can be directly used to infer gravitational potential differences between different locations as described by Müller et al. (2018). Indeed, a relative accuracy of clocks of 10^{-18} is equivalent to measuring the gravitational red shift effect between two clocks corresponding to a height difference amounting to 1 cm. Applying this conceptually new approach can facilitate or even fully resolve the problem of the systematic discrepancies existing between various height systems on the regional, national, and intercontinental levels (known to reach centimeters to decimeters). Müller et al. (2018) conclude that atomic clocks and time metrology will provide us with unique opportunities to support and largely extend the tools for gravity field measurements and physical height determination based on the direct access to the gravity potential differences.

4 Role of Clocks in Fundamental Astronomy

Timing has always been key for astronomers. The issue of data timing is fundamental for observations, whatever wavelength (visible, ultraviolet, infrared . . .) they are at. This is particularly true for the ongoing mission Gaia. With the objective of providing a highly stable and accurate time tag for each observation, the Gaia spacecraft has a Rubidium clock onboard, which is regularly synchronized with the aid of clocks on Earth. The procedure is described by Klioner et al. (2018), taking into account all known effects (e.g., relativity, tropospheric delay, etc.).

Furthermore, optical detection of molecules with large-dish ground-based telescopes or the detection with satellite-based spectrometers need accurate clocks as well. In particular, it is important for searching for variation of fundamental constants from astronomical observation of molecules, which is the subject of the paper by Ubachs (2018). The status of searches for varying constants from astronomical observation of molecules is reviewed, focusing on the dimensionless proton/electron mass ratio. Improvements, currently established on some telescope-based spectrographs, focus on the use of frequency comb lasers (a laser source whose spectrum consists of a series of discrete, equally spaced frequency lines) at the observatories. Frequency comb lasers deliver a very dense spectrum of frequency markers at a separation governed exactly by an atomic clock, thus constituting an ideal reference spectrum.

Particular attention is given to pulsar timing, as pulsars are natural cosmic clocks. On long timescales, they could rival the precision of terrestrial atomic clocks. Since pulsars provide a time-standard that is completely independently from the conditions on Earth or the Solar System as a whole, they serve as an interesting complementary tool to atomic clocks. Efforts are indeed on-going to establish such an independent time standard based on pulsars. At the same time, the clockwise behavior of pulsars also serves as an invaluable tool for precision tests of gravity under strong-field conditions and for strongly self-gravitating bodies. Here, binary pulsars test complement information available from gravitational wave detectors, and also provide stringent constraints on alternative theories of gravity. In the near future, an array of observed pulsars will allow the detection of low-frequency gravitational waves, extending the observable frequency range down to nHz-frequencies. Finally, using measurement of pulse arrival times also allows a number of applications for navigation and gravitational wave detection as described by Becker et al. (2018).

5 Role of Clocks in Planetary Science

For planetary science, accurate clocks are mainly used as part of an onboard radioscience transponder, either onboard orbiters or rovers or landers at the surfaces of planets or moons. There are several ways of implementing the radio link between a spacecraft and the Earth or between two spacecraft, as discussed by Dehant et al. (2018). In practice, one can do a direct one-way link or a two-way link (outward and return) or even three-way link (involving a triangular link). The two-way radio link is the dominant type for planetary radioscience, as an accurate spacecraft clock is not necessary onboard the spacecraft since the measurements can be calibrated using high-precision clocks on Earth. In the case of one-way radio data, however, the accuracy of the measurements depends on the accuracy of the onboard clock. For reasons of mass and power saving, the satellite clock is usually a simple ultra-stable oscillator, and thus, large equipments required for atomic clocks are not necessary.

However, the possibility of using high-performance clocks with accuracy of 10^{-14} or better in space may change the picture significantly. In particular, the special case of using laser ranging and other techniques to the Moon and the implication of having an accurate clock on the Moon is shown as an example of high-level science by Dehant et al. (2018).

6 Perspectives and Conclusion

2018 will see changes in the definition of units, which, in due course, will all rely on time and frequency standards, as time is by far the most accurately measured fundamental physical quantity. Without a doubt, future science and many applications on Earth will rely more and more on highly accurate atomic clocks. Complementary future advances in space science like geodesy, astronomy, and planetary science will follow suit, with the first and probably most demanding application being fundamental physics on ground and in space.

Space Sci Rev (2017) 212:1385–1421
DOI 10.1007/s11214-017-0361-9

Clocks in Space for Tests of Fundamental Physics

Pacôme Delva[1] · Aurélien Hees[2,3] · Peter Wolf[1]

Received: 6 December 2016 / Accepted: 29 March 2017 / Published online: 12 June 2017

Abstract In this paper we describe some of the past, present and near future experiments that take advantage of clocks in space for tests of fundamental physics, together with some of the theoretical background. It is impossible to describe all missions and projects in the field, and we will only mention some of the most important ones. Nonetheless, we hope that the reader will learn from the past, appreciate the present and look forward to the future of space clocks in fundamental physics.

Keywords Gravitational theories · Clock tests · General relativity

1 Introduction

In our present understanding, and at its most fundamental level, physics is based on two theories: the Standard Model of particle physics (SM) that describes electromagnetism and the (strong and weak) nuclear interactions, and General Relativity (GR) that accounts for all gravitational phenomena. In spite of the overwhelming success of these two theories in describing much of the observed universe, a number of open issues, both theoretical and experimental remain. The assumed validity of GR at cosmological scales, together with a hypothesis of homogeneity and isotropy, have led to the "concordance model" of cosmology, referred to as the ΛCDM model, which is in agreement with all present-day observations at large scales, notably the most recent observations of the anisotropies of the cosmic mi-

High Performance Clocks with Special Emphasis on Geodesy and Geophysics and Applications to Other Bodies of the Solar System
Edited by Rafael Rodrigo, Véronique Dehant, Leonid Gurvits, Michael Kramer, Ryan Park, Peter Wolf and John Zarnecki

✉ P. Wolf
peter.wolf@obspm.fr

1 SYRTE, Observatoire de Paris, PSL Research University, CNRS, Sorbonne Universités, UPMC Univ. Paris 06, LNE, 75014 Paris, France

2 Department of Mathematics, Rhodes University, 6140 Grahamstown, South Africa

3 Department of Physics and Astronomy, University of California, Los Angeles, CA 90095, USA

crowave background by the Planck satellite. However, important difficulties remain, in particular the necessary introduction of dark energy, described by a cosmological constant Λ, and of cold dark matter, made of some unknown, stable particle, which is not accounted for in the SM.

On the theoretical side, the SM is a quantum field theory, whilst GR, as well as many other alternative theories of gravitation, are classical. As such, they are fundamentally incomplete, because they do not include quantum effects. Most physicists agree that GR and SM are only low-energy approximations of a more fundamental theory that remains to be discovered, and that would provide a unified description of all interactions. Most attempts at such a unified theory lead to tiny violations of the basic principles of GR and/or the SM, in particular the Einstein Equivalence Principle (EEP), at a, in general unknown, level of accuracy. It is the aim of high-accuracy fundamental physics experiments, like the ones described here, to search for first experimental hints of such modifications by making use of the outstanding performance provided by modern time/frequency metrology.

Indeed, of all quantities in the SI system of units, time is the one that can be measured with by far the smallest uncertainty in relative value. Best present day clocks reach relative uncertainties in the low 10^{-18} range and progress is rapid with no obvious hard limit in sight. One of the first and most challenging applications of such clocks are the experimental study of fundamental physics, and in particular the two fundamental theories mentioned above (GR and SM). As in GR space–time and gravitation are intimately related, it is no surprise that such high performance clocks, when compared over large distances and in varying gravitational fields, are one of the prime tools in the study of gravitation, GR and fundamental physics in general. However, large distances and large variations of gravitational fields are the realm of space science, and therefore putting high performance clocks into space, whilst still a technological challenge, certainly promises highly rewarding results in terms of fundamental physics.

In this paper we describe some of the past, present and near future experiments that take advantage of clocks in space. It is impossible to describe all missions and projects in the field, and we will only mention some of the most important ones. Nonetheless, we hope that the reader will learn from the past, appreciate the present and look forward to the future of space clocks in fundamental physics.

2 Theoretical Background

The classical theory of General Relativity provides a geometrical description of the gravitational interaction. It is based on two fundamental principles: (i) the Einstein Equivalence Principle (EEP) and (ii) the Einstein field equations that can be derived from the Einstein-Hilbert action. All GR extensions or alternative theories of gravitation will break at least one of these principles.

2.1 The Einstein Equivalence Principle

The first part of GR, the EEP, published in 1911 by A. Einstein (1911), gives gravitation a geometric nature. This principle implies that gravity can be identified to space–time geometry which is described mathematically by a symmetric order 2 tensor, the space–time metric $g_{\mu\nu}$. More precisely, the EEP implies that there exists only one space–time metric to which all matter minimally couples to (see Thorne et al. 1971). In practice, this means that the equations of motion for matter can be derived from the action of the Standard Model of particles in which the Minkowski metric $\eta_{\mu\nu}$ is replaced by the space–time metric $g_{\mu\nu}$:

$$S_{\text{mat}} = \int d^4x \sqrt{-g} L_{\text{mat}}(g_{\mu\nu}, \psi) \tag{1}$$

where g is the determinant of the space–time metric and $L_{\text{mat}}(g_{\mu\nu}, \psi)$ is the standard Lagrangian for matter depending on the matter fields ψ.

From a theoretical point of view, this part of Einstein's theory allows one to derive the effects of gravitation from the space–time curvature. In particular, this implies that test bodies follow geodesics of this space–time. Furthermore, ideal clocks will measure the quadratic invariant of the space–time metric $d\tau^2 = -g_{\mu\nu}dx^\mu dx^\nu$. Similarly, the propagation of electromagnetic waves is governed by Maxwell equations in which standard derivatives are replaced by covariant derivatives. At the geometric optics approximation, this implies that light rays are described by null geodesics.

From a phenomenological point of view, three aspects of the EEP can be tested (see Will 1993, 2014): (i) the Universality of Free Fall (UFF), (ii) the Local Lorentz Invariance (LLI) and (iii) the Local Position Invariance (LPI). Furthermore, the Schiff conjecture stipulates that any complete, self-consistent theory of gravity that embodies the UFF necessarily embodies the two other parts of the EEP: the LPI and the LLI (see e.g. Will 2014). However, a precise and general quantitative relationship between the sub-principles is unknown. It depends on the underlying model and varies between different candidate theories. Furthermore, no complete proof of this conjecture has been proposed and some counter-examples are known (see the discussion in Sect. 2.1.1 of Will 2014).

2.1.1 The Universality of Free Fall

The UFF states that the motion of a test body is independent of its composition. It has been constrained with various experiments. Precise tests of the UFF compare the free fall accelerations, a_1 and a_2, of two different test bodies 1 and 2 falling in the gravitational field sourced by a body S. A succinct expression for the test of the UFF takes the form (see Will 1993, 2014)

$$\left(\frac{\Delta a}{a}\right)_{S;1-2} = 2\frac{a_1 - a_2}{a_1 + a_2} \approx \left(\frac{m_P}{m_A}\right)_1 - \left(\frac{m_P}{m_A}\right)_2 \tag{2}$$

where m_P and m_A are the passive and active masses of each body. The UFF has been tested in various cases. First, it has been tested by comparing the motion of two different macroscopic bodies of standard matter. The best current constraints on the relative differential acceleration of two bodies are at the level of 10^{-13} and have been obtained by using Lunar Laser Ranging observations (see Williams et al. 2009, 2012) and using torsion balances (see Schlamminger et al. 2008; Adelberger et al. 2009). More recently, the UFF has been tested by comparing the acceleration measured by a macroscopic mass with the acceleration measured by a microscopic, quantum system and also by comparing the accelerations measured by two different microscopic quantum systems. This has been achieved using atom interferometry. The current constraints on the UFF between macroscopic and microscopic systems are at the level of 10^{-9} (see Peters et al. 1999). Similar constraints using two different types of atom interferometers have provided constraints at the level of 10^{-9} as well (see Peters et al. 2001; Merlet et al. 2010; Zhou et al. 2015). Furthermore, there is a priori no reason that bodies with different spins react to gravity in a similar way. Therefore, the UFF has also been tested by considering atoms with different spins and the related constraints are at the level of 10^{-7} (see Tarallo et al. 2014). One can wonder if anti-matter falls any differently than standard matter. Tests of the UFF using anti-hydrogen atoms are currently on-going at CERN with the AEgIS (see Aghion et al. 2013) and GBAR (see Perez and Sacquin 2012) experiments.

Even though the existing phenomenological tests of the UFF are quite extensive for standard matter, the question of the validity of the EEP in the dark sector remains open. Since dark matter has not been directly detected so far, tests of the EEP in the dark sector are model dependent. A lot of models introduced a long range interaction in the dark sector usually leading to a fifth force experienced either by standard matter or by Dark Matter (see e.g. Frieman and Gradwohl 1991; Gradwohl and Frieman 1992; Carroll et al. 2009, 2010). This hypothetical fifth force has been constrained using various astrophysical observations (see Kesden and Kamionkowski 2006a, 2006b; Carroll et al. 2009, 2010; Stubbs 1993; Bai et al. 2015). More recently, a model of Dark Matter coupled to a scalar field has been proposed by Damour et al. (1990). This model leads to a violation of the EEP between Dark Matter and standard matter. Initially introduced as a model leading to a time variation of the gravitational constant G, its cosmological implications in the context of late-time cosmology have been thoroughly studied (see Alimi and Füzfa 2007, 2008; Füzfa and Alimi 2007). In particular, it has been shown that the violation of the EEP in the Dark sector can naturally produce Dark Energy. This violation of the EEP has recently been constrained using galactic observations by Mohapi et al. (2016).

The search for a violation of the UFF will be pursued in the future. In particular, the MICROSCOPE satellite, launched successfully on April 25th 2016 is expected to test this principle at the level of 10^{-15}. Several projects aiming at improving current tests have also been proposed.

2.1.2 Local Lorentz Invariance

Local Lorentz Invariance (LLI) is a feature of relativity stating that the outcome of any local non-gravitational experiment is independent of the velocity and orientation of the apparatus (see e.g. Will 2014). While completely integrated into special relativity, the effort to test LLI has recently increased. Indeed, it is often believed that models of quantum gravity will produce violations of Lorentz invariance. These models usually introduce a fundamental length (the Planck length). Since this length is not Lorentz invariant, in most scenarios, a breaking of Lorentz symmetry at some level is produced (see the discussion in Sect. 2.1.2 from Will 2014 or Mattingly 2005). Several formalisms have been used to test Lorentz symmetry. In the simplest one, one is looking for anisotropies in the speed of light, often parametrized by the c^2-formalism. This kind of test is usually performed using improved Michelson-like experiments (see e.g. Brillet and Hall 1979). More recent tests constrain the isotropy of atomic energy levels with a very impressive accuracy (see Lamoreaux et al. 1986).

Another widely used formalism to test Lorentz invariance is called the Robertson–Mansouri–Sexl (RMS) framework (see Robertson 1949 and Mansouri and Sexl 1977) in which a modification of the kinematic Lorentz transformation is parametrized. Deviations within this formalism are usually searched (see the discussion IV.B. of Turyshev 2009) using Michelson–Morley experiment (see e.g. Stanwix et al. 2006), Kennedy–Thorndike experiment (see Kennedy and Thorndike 1932; Wolf et al. 2003) and Ives–Stilwell experiment (see Ives and Stilwell 1938; Saathoff et al. 2003). It is interesting to notice that such a test has also been performed using GPS measurements (see Wolf and Petit 1997). All modern versions of these experiments use atomic spectroscopy, atomic clocks or highly stable oscillators.

The previous formalisms consider mainly the electromagnetic interaction. More recently, a very wide formalism has been developed to consider hypothetical violation of Lorentz symmetry in all fields of physics. This framework has been named *Standard Model Extension* (SME) and contains an impressive number of parameters encoding hypothetical deviations from the Standard Model of particles or from GR (see e.g. Colladay and Kostelecky 1997, 1998; Kostelecky and Mewes 2002 or Tasson 2014). A number of these SME coefficients have been constrained by various experiments (for a review of the current constraints on SME parameters, see Kostelecky and Russel 2011). Let us mention that in this framework, atomic clocks have been successfully used (see e.g. Wolf et al. 2006a, 2006b; Reinhardt et al. 2007; Hohensee et al. 2011).

2.1.3 Local Position Invariance

Local Position Invariance (LPI) stipulates that the outcome of any local non-gravitational experiment is independent of the space–time position of the freely-falling reference frame in which it is performed (see e.g. Will 2014). This principle is mainly tested by two types of experiments: (i) search for variations in the constants of Nature and (ii) redshift tests.

The question of the constancy of the constants of Nature was first addressed by Dirac. This question is driven by the principle of reason: what could be the reasons behind the specific values of the constants of Physics? (see the discussion in Sect. 2 of Damour 2012) This argument leads to many developments of extensions of physics where the constants of physics become dynamical entities. In parallel, many observational investigations try and search for any space/time evolution of the constants of Physics (see Uzan 2011).

Amongst all the observations performed, atomic clocks have an important role leading to currently some of the best constraints currently available. In particular, linear drifts in the evolution of the fine structure constant α, in the ratio μ between the mass of the electron and the mass of the proton and in the ratio between the mass of the light quarks (up and down) and the quantum chromodynamics (QCD) energy scale Λ_{QCD}. Several groups in the world have pursued effort to constrain such hypothetical linear drifts: at SYRTE, Observatoire de Paris (see Guéna et al. 2012), NIST (see Rosenband et al. 2008), Berkeley (see Leefer et al. 2013), NPL (see Godun et al. 2014), PTB (see Huntemann et al. 2014),.... The current constraints on the variation of the three constants (fine structure constant, ratio between the mass of the electron and the mass of the proton and ration between the mass of the light quarks and the QCD energy scale) are at the level of 10^{-16} per year. More recently, searches for a harmonic temporal variation of the constants of Nature using atomic clocks have been performed in Berkeley (see Van Tilburg et al. 2015) and at SYRTE (see Hees et al. 2016a).

In addition to atomic clocks, several astrophysical observations have also been used to search for a temporal evolution of the constants of Nature. In particular, observations of quasar absorption lines at the Very Large Telescope (VLT) in Chile and at the Keck Observatory (Hawaii) lead to observations of $\Delta\alpha/\alpha$ of the order of 10^{-6} at cosmological scales, typically for redshifts between 0.2 and 4 (see Webb et al. 2003; Srianand et al. 2004). It is interesting to mention that both sets of observations seem to indicate a temporal evolution of the fine structure constant but the two estimations are not compatible with each other. This led the two teams to suggest a dipolar dependence of the fine structure constant in the sky (see Webb et al. 2011). Using a linear interpolation at low redshift, we can compare the order of magnitude of constraints obtained with atomic clocks and using quasar observations. The atomic clock estimations are roughly one order of magnitude more accurate than quasar observations. This comparison implies a linear approximation of the cosmological evolution and implies that no screening mechanism is playing a role around the Earth (a screening mechanism can lead to a situation where deviations from GR at cosmological scales are hidden or strongly reduced on local scales, see for example Khoury and Weltman 2004a, 2004b). Furthermore, observations of the Cosmic Microwave Background (CMB) led to a measurement of $\Delta\alpha/\alpha$ at the level of 10^{-3} for a redshift $z \approx 1000$ (see Planck Collaboration 2013). Finally, analysis of the Big Bang Nucleosynthesis (BBN) also led to an estimation of $\Delta\alpha/\alpha$ at the level of 10^{-3} for a redshift z of the order of 10^{10} (see Avelino et al. 2001).

In addition to temporal variations of the constants of Nature, one can search for spatial variations. Regarding this, atomic clocks have also been widely used to search for a variation of the constants of Nature with respect to the gravitational potential of the Sun. The idea is to compare two clocks working on different atomic transitions and therefore sensitive differently to the different constants of Physics, located at the same place and search for periodic variations in their frequencies comparison. This kind of test is also known as null redshift experiment (see Will 2014). Several groups in the world have been measuring this effect which is now constrained at the level of 10^{-6}: the SYRTE (see Guéna et al. 2012), the USNO (see Peil et al. 2013), Berkeley (see Leefer et al. 2013), NIST (see Ashby et al. 2007b), ...

The second way to test the LPI is to measure the gravitational redshift. The gravitational redshift is a consequence of the EEP predicted by A. Einstein (1911). It was observed for the first time by Pound and Rebka (1959). A simple and convenient formalism to test the gravitational redshift is to introduce a new parameter α_{red} defined through (see Will 2014)

$$\frac{\Delta\nu}{\nu} = (1 + \alpha_{\text{red}})\frac{\Delta U}{c^2} \tag{3}$$

with α_{red} vanishing when the EEP is valid. The best constraint on α_{red} is at the level of 10^{-4} and has been obtained in 1976 by comparing the frequency of two clocks: one onboard a rocket and the other one on Earth (see Vessot et al. 1980). Improvements on this test is expected in the near future by using the GNSS satellites Galileo V and VI (see Delva et al. 2015 and Sect. 3 below) and with the Atomic Clock Ensemble in Space project (see Cacciapuoti and Salomon 2009 and Sect. 3 below) that aims at comparing atomic clocks on the international space station with atomic clocks on Earth. Finally, a test of the gravitational redshift is expected in the near future using spectroscopy observations of S stars orbiting around our galactic center Sagitarius A* (see Zucker et al. 2006).

2.1.4 Summary

In conclusion, the EEP is an essential part of Einstein's theory that is shared by a number of alternative theories of gravitation. All the theories of gravitation satisfying the EEP are called *metric theories*. In these theories, it is sufficient to know the space–time metric (to which matter is minimally coupled) to infer all effects produced by gravitation (motion of bodies, light propagation, behavior of clocks,...). It is interesting to mention that the equivalence principle is a generalization of the fact that all bodies seem to fall with the same acceleration in a gravitational field. Nevertheless, as mentioned by T. Damour (2012): *"Despite its name, the 'equivalence principle' (EP) is not one of the basic principles of physics. There is nothing taboo about having an observable violation of the EP. In contrast, one can argue (notably on the basis of the central message of Einstein's theory of general relativity) that the historical tendency of physics is to discard any, a priori given, absolute structure (principle of absence of absolute structures).*" It is therefore highly important to pursue our quest to test the various facets of this principle. Several on-going projects or proposals aim at improving our current searches for deviations from the EEP. In all the different kind of tests, clocks have an important role regarding tests of the LLI and of the LPI through redshift tests and tests of the constancy of the constants of Nature.

2.2 Einstein Field Equations

The EEP allows one to derive the effects of gravitation from the space–time metric $g_{\mu\nu}$. In order to have a complete theory, one needs to give a prescription to determine the form of this metric. In GR, the form of the metric is determined by solving the Einstein field equations

$$R_{\mu\nu} - \frac{1}{2} R g_{\mu\nu} + \Lambda g_{\mu\nu} = \frac{8\pi G}{c^4} T_{\mu\nu}, \tag{4}$$

where $R_{\mu\nu}$ and R are the Ricci tensor and scalar curvature from the metric, G the gravitational constant, c the speed of light in vacuum, $T_{\mu\nu}$ the stress-energy tensor and Λ the cosmological constant. These equations are making a very interesting link between space–time geometry essentially described in the left part of the equation and the matter/energy content in this space–time essentially described by the stress-energy tensor. In other words, this set of equations describes how space–time is curved by the presence of matter and energy.

The Einstein field equations can be derived by means of a variational principle from the Einstein-Hilbert action (see Einstein 1916)

$$S_{\mathrm{grav}} = \frac{c^4}{16\pi G} \int d^4x \sqrt{-g} (R - 2\Lambda). \tag{5}$$

The total action which describes completely the GR theory is the sum $S_{\mathrm{grav}} + S_{\mathrm{mat}}$ which makes GR an extremely simple theory from a conceptual point of view. It is also interesting to mention that due to the Lovelock theorem, the Einstein field equations are the only second order equations that can be derived from a least-principle action based on the space–time metric (and its derivatives) in a 4 dimensional Riemannian space (see Lovelock 1971, 1972). Following this theorem, alternative (metric) theories of gravitation will always imply one of the following (see Clifton et al. 2012):

- The existence of new fields in addition to (or instead of) the space–time metric
- The existence of higher order derivatives of the metric in the field equations
- To work in a space–time with higher dimension than 4
- To give up locality.

If the EEP specifies how different types of mass-energy react to gravitation, the Einstein field equations govern how gravitation is generated by matter and energy. A modification of the Einstein-Hilbert action will lead to different field equations that may lead to differences in the space–time metric for a given situation. From an experimental point of view, it is of prime importance to search for metric deviations that would be produced by any deviations from GR. The class of metric theories of gravitation is huge and it is particularly difficult to analyze observational data in a general framework. So far, two frameworks have been widely used to analyze data: the parametrized post-Newtonian (PPN) formalism (see Will 1993, 2014) and the fifth force formalism (see Talmadge et al. 1988;

Fischbach and Talmadge 1999; Adelberger et al. 2009). As we shall see in the coming sections, these two formalisms have been very precisely constrained with observations at Solar System scales. Nevertheless, it is useful to analyze experimental results in new extended frameworks. Indeed, even if observations lie very close to GR when analyzed within the PPN or fifth force framework, this does not mean that this has to be true in any other framework. The existing frameworks indeed cover a limited set of alternative theories of gravity (see Hees et al. 2012, 2014a). For example, a formalism aiming at considering violation of Lorentz invariance in the gravity sector has been develop within the SME framework (see Bailey and Kostelecky 2006). In this formalism, an expansion at the level of the action is performed which naturally leads to a post-Newtonian metric that differs from the PPN one. Other examples of theories not entering the PPN or fifth force frameworks are given by the MOND (MOdified Newtonian Dynamics) phenomenology which produces a quadrupolar deviation from the Newton potential in the Solar System (see Milgrom 2009; Blanchet and Novak 2011), by the post-Einsteinian Gravity (PEG, see Jaekel and Reynaud 2005, 2006), by the parametrized post-Newtonian–Vainshteinian formalism (see Avilez-Lopez et al. 2015), by Horndeski's gravity (see Hohmann 2015), etc. Currently, only few data analyses are exploring these formalisms or theories.

2.2.1 *The Parametrized Post-Newtonian (PPN) Formalism*

The PPN formalism is a nice interface between theoretical developments and data analyses. It is fully described in Will (1993). In this formalism, a phenomenological expansion at the level of the space–time metric is performed by introducing 10 dimensionless parameters. From an observational point of view, these parameters can be constrained regardless of any considerations about the hypothetical underlying theory. From a theoretical point of view, if the post-Newtonian metric of a theory can be matched to the PPN metric, it can automatically be constrained by all observations used to constrain the PPN parameters.

In the simplest case, only two PPN parameters are considered and the space–time metric for a spherically symmetric configuration is given by

$$ds^2 = -\left(1 + 2\frac{\phi_N}{c^2} + 2\beta\frac{\phi_N^2}{c^4} + \cdots\right)c^2dt^2 + \left(1 - 2\gamma\frac{\phi_N}{c^2} + \cdots\right)dx^2, \tag{6}$$

where ϕ_N is the Newtonian potential and γ and β are the PPN parameters. These parameters take the value of 1 in GR and may deviate from 1 in alternative theories.

The PPN formalism was historically the first one developed and therefore was extensively constrained by several different types of observations. The best constraint on the γ parameter was obtained by measuring precisely the Shapiro time delay with the Cassini spacecraft when it was cruising between Jupiter and Saturn in 2003 and is given by (see Bertotti et al. 2003)

$$\gamma - 1 = (2.1 \pm 2.3) \times 10^{-5}. \tag{7}$$

This constraint has been confirmed by measuring the light deflection around the Sun using Very Long Baseline Interferometry (see Lambert and Le Poncin-Lafitte 2011), by using radioscience observations of spacecraft orbiting Mars (see Konopliv et al. 2011) or Mercury (see Verma et al. 2014).

The β PPN parameter is essentially constrained through orbital dynamics. Lunar Laser Ranging analysis provides a constraint at the level of 10^{-4} (see Williams et al. 2009, 2012). Furthermore, planetary ephemerides analysis currently provides constraints at the level of

10^{-5} (see Konopliv et al. 2011 for JPL ephemerides, Fienga et al. 2015 for the INPOP ephemerides and Pitjeva and Pitjev 2013 for the EPM ephemerides).

The other PPN parameters are also very well constrained. An updated list of constraints can be found in Will (2014). Several projects will improve current constraints in the future like e.g. the GAIA space mission (see Mignard and Klioner 2010; Mouret 2011; Hees et al. 2015a) or BepiColombo (see Milani et al. 2002; Ashby et al. 2007a). Furthermore, several proposals aiming at improving PPN constraints by more than one order of magnitude have been proposed (see Sect. 4 below).

2.2.2 *The Fifth Force Formalism*

In the fifth force formalism, a Yukawa-type deviation from Newtonian gravity is considered. The deviation from the Newtonian potential is given by

$$\phi = -\frac{GM}{r}\left(1+\alpha e^{-\frac{r}{\lambda}}\right), \tag{8}$$

where α is the strength of the interaction and λ a characteristic length scale. The idea is to constrain the couple of parameters (α, λ). An impressive number of experiments and observations have also been used to constrain this formalism at various scales: at Solar System scales by using planetary ephemerides analyses (see Konopliv et al. 2011), Lunar Laser Ranging observations (see Adelberger et al. 2003), ranging measurements of LAGEOS satellite around Earth (see Fischbach and Talmadge 1999). At lower distances, several geophysical observations have been used to constrain a hypothetical fifth force. A short distances, torsion balances provide the best constraints. For $\lambda \leq \mu m$, it becomes extremely hard to constrain the strength of the interaction because of several quantum noise sources (like e.g. the Casimir effect). In conclusion, the fifth force formalism is very well constrained except at very low and very large distances where deviations can still be searched for.

2.2.3 *The Pure Gravity Sector of the Standard Model Extension*

As mentioned previously, the SME is a very wide framework aiming at considering systematically Lorentz violations in any sector of Physics. In the pure gravity sector, SME can be interpreted as a metric extension of GR. This framework is built by considering an expansion at the level of the action by introducing terms that violate Lorentz symmetry (see Bailey and Kostelecky 2006). In the minimal SME framework, the resulting post-Newtonian metric is parametrized by a symmetric trace-free tensor $\bar{s}^{\mu\nu}$ (i.e. by 9 coefficients). This metric does not correspond to the PPN metric and therefore, this framework offers new possibilities to search for GR deviations. So far, a couple of postfit data analyses have been performed to constrain the SME coefficients. A postfit analysis means that the analysis has been done in two steps: (i) a standard reduction of the data in GR followed by (ii) a fit of SME signatures in the residuals resulting from the GR reduction. This approach is quite simple to implement in practice but suffers from the fact that all correlations between the SME coefficients and other standard parameters fitted in the analysis is neglected and the obtained uncertainties on the estimated SME parameters are therefore usually underestimated. Therefore, these analyses should be considered as sensitivity analyses only. Sensitivity analyses have been performed with LLR observations (see Battat et al. 2007), atom interferometry measurements (see Müller et al. 2008), binary pulsars observations (see Shao 2014), planetary ephemerides analyses (see Iorio 2012; Hees et al. 2015b) and Gravity Probe B analysis (see Bailey et al. 2013).

In addition, very recent analyses of VLBI data and of LLR data have been performed within the SME framework (see Le Poncin-Lafitte et al. 2016; Bourgoin et al. 2016). Contrarily to previous analyses, these studies are not post-fit analyses. Rather, the SME contribution has been added to the full software used to analyze VLBI or LLR data.

2.2.4 *Modified Newtonian Dynamics (MOND)*

The MOND phenomenology was introduced in 1983 by M. Milgrom (1983a, 1983b, 1983c) in order to explain observations at the galactic scale (in particular galactic rotation curves) without the introduction of Dark Matter. The main idea behind the MOND phenomenology is to replace the standard Newton acceleration g_N valid in "high" gravitational field by $\sqrt{a_0 g_N}$ for regions of the Universe where the gravitational field is very low. More precisely, the MOND phenomenology depends on a MOND interpolating function ν and on a MOND acceleration scale a_0. The interpolating function is making the transition between the Newtonian regime and the MONDian regime appearing in low gravitational fields. While developed initially as a purely phenomenological Newtonian model, the MOND phenomenology has later been extended as relativistic theories. A review of different relativistic MOND theories can be found in Bruneton and Esposito-Farèse (2007) or in Famaey and McGaugh (2012). A review of the different interpolating functions used in the literature can also be found in Famaey and McGaugh (2012) (see also Hees et al. 2016b).

Recently, it has been shown that contrarily to what was previously expected, the MOND phenomenology would produce detectable effects in the Solar System. This effect is due to the non-linearity of the MOND phenomenology which implies that the external galactic gravitational field will play a role in this theory. This External Field Effect was first discovered by Milgrom (2009) and was studied later by Blanchet and Novak (2011). The main contribution from this External Field Effect in the Solar System takes the form of a quadrupolar deviation from the Newton potential

$$\phi_N = -\frac{GM}{r} - \frac{Q_2}{2} x^i x^j \left(e_i e_j - \frac{1}{3}\delta_{ij} \right), \tag{9}$$

where the unit vector e_i points toward the galactic center and Q_2 is a parameter that depends directly on the MOND interpolating function ν and on the MOND acceleration scale a_0. Values of this parameter for different interpolating functions are given in Milgrom (2009), Blanchet and Novak (2011) and Hees et al. (2016b).

The interesting point is that this modification of gravitation does not enter the fifth force or the PPN formalisms and it can be used efficiently to detect or constrain the MOND phenomenology. A recent analysis of radio tracking data of the Cassini spacecraft when it was orbiting Saturn (between 2004 and 2013) led to a constraint on Q_2 given by (see Hees et al. 2014b)

$$Q_2 = (3 \pm 3) \times 10^{-27}\ \text{s}^{-2}. \tag{10}$$

This constraint on MOND theory is the best currently available on Solar System scales and the propositions to use LISA Pathfinder to detect MOND theory close to a saddle point (see e.g. Bekenstein and Magueijo 2006) are unlikely to improve this test (see Hees et al. 2016b).

2.2.5 *Summary*

The Einstein Field Equations determine the form of the metric, or in other words describe how gravitation is generated. They can be derived from the Einstein–Hilbert action which gives GR a very simple form (at least conceptually). Many alternatives theories of gravitation have been developed introducing new additional fields, including higher order derivative terms of the metric in the field equations (like e.g. $f(R)$ gravity), working in higher dimensional space–time, ... These modifications of gravitation are usually motivated by developments of quantum theory of gravity, the unification of the interactions of physics, hypothetical models of Dark Energy and Dark Matter.

From an observational point of view, it is difficult to constrain deviations from GR in a single general framework. Nevertheless, so far, two frameworks have been widely used to search for GR deviations: the PPN formalism and the fifth force framework. As mentioned in this section, a lot of measurements have been used to constrain these, and most of them rely on high performance clocks either directly or indirectly (e.g. through deep space radio ranging). It is therefore interesting to consider other alternative frameworks that allow to search for deviations that may not be seen within the standard formalisms. The gravity sector of SME and the MOND External Field Effect developed above are two examples of these alternative frameworks that have recently been constrained with data but many other examples exist in the literature.

2.3 Link Between Local Tests of GR and Galactic and Cosmological Observations

The previous section was devoted to describe briefly the theoretical basis of GR and how these can be tested experimentally at small scales (Solar System and smaller). Here, we will show how the interplay between local scales and large scales (i.e. galactic and cosmological scales) can be highly beneficious. First of all, a large class of alternative theories of gravitation have been developed to specifically tackle the Dark Matter and/or Dark Energy problematics. The idea is to modify the gravitation theory to explain observations that require the introduction of Dark Matter and/or Dark Energy without adding new components to the standard model of particles or introducing the cosmological constant. An excellent review of a large class of theories developed for cosmological purposes can be found in Clifton et al. (2012). As can be seen from this review (and the list is not exhaustive), the number of theoretical models developed is huge. It is therefore of prime importance to confront these GR extensions intended for large scales, with small scale observations to try and identify the theories that are viable on all scales in our Universe. The idea is therefore to combine observations at different scales to reduce as much as possible the compatible parameters space. In the next section, we will develop several examples to illustrate this point and to highlight the power of this type of combined analysis. Ultimately, one would try to identify a theory (or set of theories) that satisfies all current observations at all scales and that will solve some (or all if we are optimistic) problems related to the gravitation theory.

2.3.1 "Dilaton-Like" Scalar Fields

Amongst the zoo of alternative theories of gravitation, tensor-scalar theories are very popular and have been studied in great details. The pioneering work on tensor-scalar theories was performed by Jordan, Brans and Dicke (see Brans and Dicke 1961) motivated by considerations about Mach's principle. From there, an impressive number of studies of tensor-scalar theories have been performed by modifying the interaction with the space–time metric, modifying the kinetic part of the scalar field action, by considering different type of potentials, by allowing more than second order derivatives in the field equations, by considering several different couplings to matter, and so on.... The motivations to consider such models are also very wide: they are considered as phenomenological model inspired by unification theory (see e.g. Damour and Polyakov 1994), as higher dimensional theoretical models (see e.g. Overduin and Wesson 1997), as a model of Dark Matter (see e.g. Stadnik and Flambaum 2015 or Arvanitaki et al. 2015), as model of Dark Energy (see for example Khoury and Weltman 2004a, 2004b), ...

The dilaton theories are a class of models where the scalar field is non-minimally coupled to matter, therefore breaking explicitly the EEP. As an example, the coupling between matter and the scalar field can be parametrized by the interaction Lagrangian (see e.g. Damour and Donoghue 2010)

$$L_{\text{int}} = \varphi\left(\frac{d_e}{4\mu_0}F_{\mu\nu}F^{\mu\nu} - \frac{d_g\beta_g}{2g_3}G^A_{\mu\nu}G^{\mu\nu}_A - c^2\sum_{i=e,u,d}(d_{m_i} + \gamma_{m_i}d_g)m_i\overline{\psi_i}\psi_i\right), \tag{11}$$

where $F_{\mu\nu}$ is the standard electromagnetic Faraday tensor, μ_0 the vacuum magnetic permeability, $G^A_{\mu\nu}$ the gauge invariant gluon strength tensor, g_3 the QCD gauge coupling, β_3 denotes the β function for the running of g_3, m_i the mass of the fermions ($i = e, u, d$ for electron and up/down quark), γ_{m_i} the anomalous dimension giving the energy running of the masses of the QCD-coupled fermions and ψ_i the fermions spinor. This Lagrangian is parametrized by five dimensionless coefficients d_e, d_{m_e}, d_{m_u}, d_{m_d} and d_g that characterize the coupling between the scalar and standard model fields. More complex type of couplings or coupling to other part of the standard model (e.g. to gauge bosons of the weak interaction) have also been explored (see e.g. Stadnik and Flambaum 2014 and references therein). It is well known that such a model will induce a violation of the Einstein Equivalence Principle for baryonic matter (see e.g. Damour and Donoghue 2010) and can be constrained using all observations related to the EEP (see Sect. 2.1 above).

For example, in the case where the action of the scalar field is standard and no potential is considered (massless field), it is known that UFF tests constrain the d_g, d_{m_u} and d_{m_d} coefficients pretty well (see Damour and Donoghue 2010) while tests with clocks (redshift tests or tests of the constancy of the constants of Physics) can be interesting regarding the d_e and more importantly the d_{m_e} coefficients (see Hees and Minazzoli 2015). In the case where the scalar field is slightly massive, clock comparisons can be much better than UFF experiments (see Van Tilburg et al. 2015 or Hees et al. 2016a). This illustrates the complementarity of the different types of experiments. This class of theories has several links to galactic and cosmological scales. First of all, the dilaton coupling introduced in the interaction Lagrangian above has very specific signatures at the cosmological level and can impact several cosmological observations (Cosmic Microwave Background—CMB—temperature, distance duality relation, ...). Furthermore, this class of models has been studied as Dark Matter and Dark Energy model and it is therefore interesting to make a combined analysis to constrain these models. We will explore these three cases here in order to illustrate the strong interaction between the different scales.

Some Cosmological Observables In this section, we will consider a dilatonic field that couples only to EM so that the related part of the action is

$$S_{\text{EM}} = -\frac{1}{4\mu_0 c}\int d^4x\sqrt{-g}(1 - d_e\varphi)F_{\mu\nu}F^{\mu\nu} + q_p\int A_\mu dx^\mu \tag{12}$$

where $q_p$is the electric charge of a particle interacting with the EM field. It is important to point out that in order to preserve the U(1) gauge invariance, the interaction part of the electromagnetic Lagrangian cannot include a contribution from the scalar field (see Bekenstein

1982; Hees et al. 2015c). The coupling introduced breaks explicitly the EEP (in particular, it cannot be absorbed through a conformal transformation). As mentioned previously, this kind of coupling has implications on the universality of free fall and on variations of fundamental constants. In addition to these effects, four cosmological observables are modified (with respect to GR) and are intimately related to each other in this class of theories (see Hees et al. 2014c): (i) temporal variation of the fine structure constant, (ii) violation of the distance-duality relation, (iii) modification of the evolution of the CMB temperature and (iv) CMB spectral distortions. The theoretical derivation can be found in Hees et al. (2014c). It is worth to insist on the fact that the derivation relies only on the matter part of the action and not on the gravitational part. This means that our results apply to a very wide class of gravitation theories. In a Friedman–Lemaître–Robertson–Walker space–time, the expressions of the four observables are given by the following expressions:

- Temporal variation of the fine structure constant. A straightforward identification in the action leads to (see e.g. Uzan 2011) $\frac{\Delta\alpha}{\alpha} = \frac{\alpha(z)-\alpha_0}{\alpha_0} = d_e(\varphi(z) - \varphi_0)$, where z is the redshift and subscripts 0 refer to $z = 0$.
- Violation of the cosmic distance-duality relation. The optic geometric limit of the modified Maxwell equations shows that photons propagate on null geodesics but their number is not conserved due to an exchange with the scalar field (see also Minazzoli and Hees 2014). Therefore, the expression of the angular diameter distance (D_A) is the same as in GR but this leads to a modification of the distance-luminosity expression (D_L) (see also Minazzoli and Hees 2014) and hence to a violation of the cosmic distance-duality relation:

$$\eta(z) = \frac{D_L(Z)}{[D_A(z)(1+z)^2]} = 1 + \frac{d_e}{2}\big(\varphi(z) - \varphi_0\big).$$

- Modification of the evolution of the CMB temperature. Considering the CMB as a gaz of photons described by a distribution function solution of a relativistic Boltzman equation and using the geometric optic approximation of the modified Maxwell equations leads to a modification of the CMB temperature evolution:

$$T(z) = T_0(1+z)\big[1 + 0.12 d_e\big(\varphi(z) - \varphi_0\big)\big].$$

- Spectral distortion of the CMB. Using the same approach as the one sketched in the last item, one can show that the evolution of the CMB radiation leads to deviations from its black body spectrum. When this deviation is parametrized by a chemical potential μ, one can show that its expression at current epoch is given by $\mu = 0.47 d_e(\varphi(z = z_{\mathrm{CMB}}) - \varphi_0)$.

To summarize, a coupling between the scalar field and EM implies that the four observables are intimately linked to each other through the relations

$$\begin{aligned} d_e\big(\varphi(z) - \varphi_0\big) &= \frac{\Delta\alpha}{\alpha} = 2\big(\eta(z) - 1\big) = 8.33\left(\frac{T(z)}{T_0(1+z)} - 1\right)\mu \\ &= 0.47\frac{\Delta\alpha(z_{\mathrm{CMB}})}{\alpha} = 0.94\big(\eta(z_{\mathrm{CMB}}) - 1\big) = 3.92\left(\frac{T(z_{\mathrm{CMB}})}{T_0(1+z_{\mathrm{CMB}})} - 1\right). \end{aligned} \tag{13}$$

These relations hold for all theories whose matter part of the action can be cast in the form of the action (12). This class of theories is very large and includes all metric theories. These relations imply also that if a deviation of the EEP is observed around Earth, it is likely to have cosmological counterparts as well, which would be an excellent check to confirm any observed deviations.

These relations can also be used for two distinct purposes: (i) use measurements of one type of observations to constrain another type (for example, use observation of the variation of α to constrain the evolution of $T_{\rm CMB}$) and (ii) use two different types of observations and see if they are consistent with the relations above to test them.

From an observational point of view, several parameterizations of the distance-duality relationship have been considered in the literature: $\eta(z) = 1 + \eta_0$, $\eta(z) = 1 + \eta_1 z$, $\eta(z) = 1 + \eta_2 \frac{z}{1+z}$, $\eta(z) = 1 + \eta_3 \ln(1 + z)$, $\eta(z) = (1 + z)^{\varepsilon}$. On the other hand, essentially one parametrization has been used to search for deviations in the evolution of the CMB temperature $T(z) = (1 + z)^{1-\beta}$. Assuming that the theory of gravitation is described by the multiplicative coupling introduced in Eq. (12) (which is a large class of theories including GR), we can use the relations from Eq. (13) to transform observational constraints on one type of observations into constraints on another type. As an example, we use three datasets of α measurements: precise laboratory clock measurements of variations of α providing the constraint (see Rosenband et al. 2008) $\dot{\alpha}/\alpha = (-1.6 \pm 2.3) \times 10^{-17}$/yr, 154 quasar absorption lines observed at the VLT (see King et al. 2012) and 128 quasar absorption lines observed at the Keck observatory (see Murphy et al. 2003). Using separately these three datasets and Eq. (13), we constrain the parameters η_i, ε and β by using a Bayesian inversion (see Hees et al. 2014c).

Parameter	Estimation [$\times 10^{-7}$]		
	VLT	Keck	Clocks
η_0	10 ± 6	-29 ± 10	–
η_1	8.4 ± 3.5	-16 ± 6	1.0 ± 1.4
η_2	20 ± 10	-49 ± 17	1.0 ± 1.4
η_3	14 ± 6	-30 ± 11	1.0 ± 1.4
ε	14 ± 6	-30 ± 11	1.0 ± 1.4
β	-3.3 ± 1.5	7.2 ± 2.5	-0.3 ± 0.3

The constraints derived from clocks measurements rely only on one observation and is valid only if no screening mechanism occurs around Earth. The obtained constraints improve by 5 orders of magnitude direct observations of η or of the CMB temperature. A similar analysis has been performed in order to test the relations given by Eq. (13). No deviations have been observed at the level of 10%. The main limitation in this test is currently coming from angular diameter observations. The future project EUCLID or the SKA are expected to improve these by one order of magnitude.

Dilaton Field as Dark Matter Another example of interaction between laboratory and larger scales is to consider models where Dark Matter is made of a non-minimally coupled scalar field. While DM particles have been widely searched with high energy physics measurements, gravitational observations can also provide some insight in this search. For example, a model of scalar DM has been proposed (see e.g. Derevianko and Pospelov 2014; Stadnik and Flambaum 2014, 2015; Arvanitaki et al. 2015)

$$S = \frac{c^3}{16\pi G} \int d^4x \sqrt{-g} \left[R - 2g^{\mu\nu}\partial_\mu\varphi\partial_\nu\varphi - 2\frac{c^2}{\hbar^2} m_\varphi^2 \varphi^2 \right] + S_{\rm SM}[g_{\mu\nu}, \psi] + S_{\rm int}[g_{\mu\nu}, \varphi, \psi] \tag{14}$$

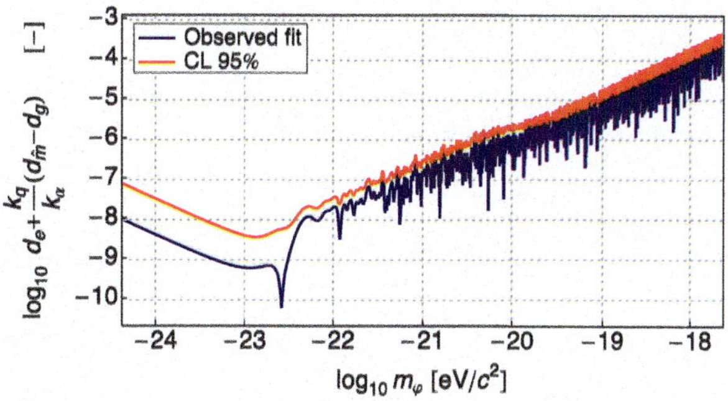

Fig. 1 Estimated values of the linear combination $d_e + 0.043(d_{\hat{m}} - g_g)$ of coupling constants d_i between a massive scalar field and standard matter fields as a function of scalar field mass m_φ. The best fit values are shown in *blue*, with the 95% confidence upper bounds in *red*. From Hees et al. (2016a)

where S_{SM} represents the Standard Model action and S_{int} is the action of the interaction between the scalar field and matter constructed from the Lagrangian given by Eq. (11). This class of models has several gravitational signatures that can be searched for at all scales of physics (see e.g. Arvanitaki et al. 2015; Stadnik and Flambaum 2015). In particular, one can show that under reasonable hypotheses, the cosmological evolution of the scalar field is harmonic: $\varphi = cst + \varphi_0[\cos c^2 m_\varphi t/\hbar + \delta]$. For low m_φ the mean cosmological pressure due to the harmonic part of the scalar field vanishes allowing to identify the scalar field as DM. This identification implies that the amplitude of the scalar field oscillation φ_0 is directly related to the DM energy density, which is observed in our galaxy as being roughly 0.4 GeV/cm^3 (see McMillan 2011). Such a model would produce oscillations in the different constants of Physics that can be searched for in comparison of atomic clocks frequency (see Arvanitaki et al. 2015; Stadnik and Flambaum 2015). These oscillations depend directly on the matter/scalar field coupling constants d_i introduced in the Lagrangian from Eq. (11). Recently, this model has been constrained by using Dysprosium (Dy) atomic transitions at Berkeley (see Van Tilburg et al. 2015) and using Cs/Rb atomic transitions at SYRTE (see Hees et al. 2016a). The best constraint on the d_i coefficients obtained using atomic clocks is shown in Fig. 1. We can exclude couplings larger than 5.3×10^{-4} at any m_φ within our range, with our most stringent limit being as low as 3.8×10^{-9} at $m_\varphi = 1.4 \times 10^{-23}$ eV$/c^2$. Although, these constraints are numerically much better than the ones obtained with UFF experiments (see Arvanitaki et al. 2016), they are complementary to those of Van Tilburg et al. (2015) and also to those coming from tests of the UFF as they probe different combinations of the coupling constants d_i.

This example shows once again how local measurement can help to constrain models constructed in order to explain large scales observations.

Dilaton Field as Dark Energy The same action as the one introduced in the previous section, but with a vanishing potential, has also been studied in the context of Dark Energy—DE (see e.g. Carroll 1998 or Nunes and Lidsey 2004 and references therein). In this case, it is useful to work with the cosmological energy density and pressure associated with the scalar field and it can be shown that the DE equation of state $w_\varphi = p_\varphi/\rho_\varphi$ is given by (see Nunes and Lidsey 2004) $w_\varphi = -1 + \frac{2}{3}\frac{\varphi'^2}{\Omega_\varphi}$ where $\Omega_\varphi = \frac{8\pi G \rho_\varphi}{3H^2c^2}$. This equation allows to derive an expression that makes the connection between the temporal evolution of α to cosmological variables (see e.g. Martins and Pinho 2015)

$$\left.\frac{\dot{\alpha}}{\alpha}\right|_0 = -d_e H_0 \sqrt{\frac{3}{2}\Omega_{\varphi 0}(1 + w_{\varphi 0})}. \tag{15}$$

Fig. 2 One-, two-, three-sigma constraints on the d_e–w_0 plane obtained by using atomic clock measurements (only *red curves*), cosmological data only (*blue lines*) and a combination of both datasets (*black curves*). From Martins and Pinho (2015)

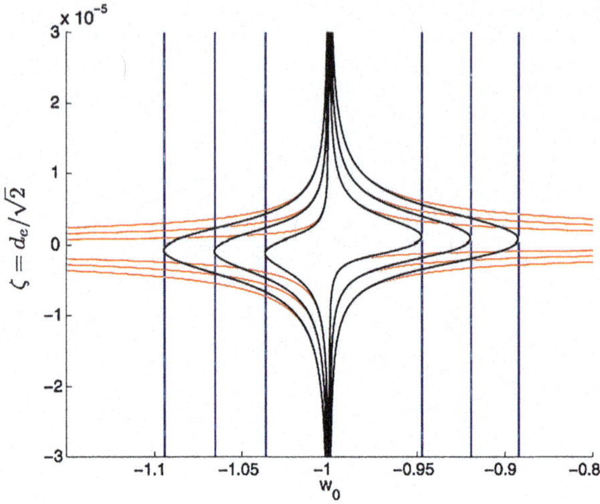

This relation is very useful in order to combine data in a global analysis. For example, Martins and Pinho (2015) uses atomic clocks constraints on $\dot{\alpha}/\alpha$ and cosmological observations (Supernovae Ia and Hubble parameter measurements) in order to constrain d_e and $w_{\varphi 0}$ (under the assumption that w_φ is constant and using a fixed value of the Hubble constant of 70 km/s/Mpc). Figure 2 represents the constraints in the $(d_e, w_{\varphi 0})$ plane obtained using atomic clock observations only, using cosmological observations only and using a combination of both datasets. It is interesting to notice that both individual datasets are sensitive to only one of the parameters. It is really the combination of local measurements with large scales observations that seriously constrains this model.

2.3.2 The Chameleon Field

The examples developed in the previous section are all based on a dilaton-like model, meaning that they all break the EEP. The interest of combining small scale measurements with large scales observations is of course not restricted to this class of theories. As an example, in this section, we will show that the chameleon fields are also interesting in this perspective.

The chameleon fields have been introduced by Khoury and Weltman (2004a, 2004b) and are the first example of what is nowadays known as "screening mechanism". A screening mechanism refers to a mechanism that hides a large deviation from GR in certain regions of space–time. The main motivation behind these mechanisms is coming from the DM and DE problematic and trying to solve them by modifying significantly the GR theory at large scales. In order to satisfy the stringent local tests of gravitation, a screening mechanism is often involved. In the particular case of the chameleon fields, the screening is based on the local matter density: in regions of space time where the matter density is large (like e.g. in the Solar System), the chameleon field is highly massive and therefore frozen and undetectable while in region of space–time where the matter density is low (i.e. at cosmological scales), the chameleon field is nearly massless and can eventually play the role of DE. The original chameleon model can be expressed as

$$S = \int d^4x\sqrt{-g}\left[\frac{m_p^2}{16\pi}R - \frac{1}{2}\partial_\mu\phi\partial^\mu\phi - V(\phi)\right] + S_{\mathrm{mat}}\left[\tilde{g}_{\mu\nu} = A^2(\phi)g_{\mu\nu}, \psi\right], \tag{16}$$

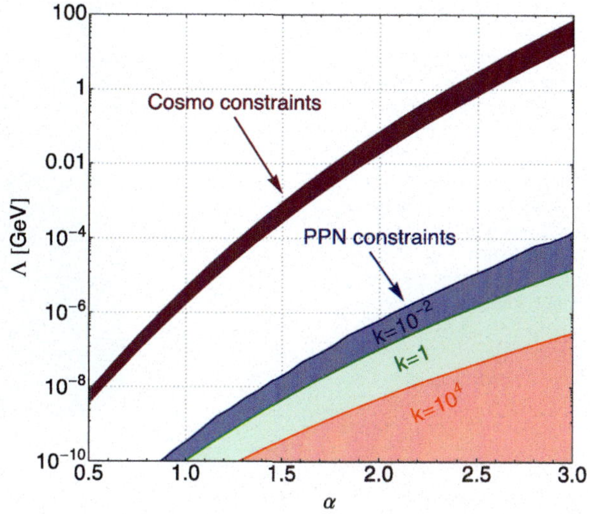

Fig. 3 Representation of the constraints in the (α, Λ) plane obtained with Supernovae Ia observations and with the Cassini PPN constraints (see Eq. (16)). The cosmological constraint is nearly insensitive to the value of k while the PPN constraint depends on the value of the coupling parameter. From Hees and Füzfa (2015)

where $m_p^2 = 1/G$ (here, we use natural units in which $\hbar = c = 1$) and $A(\phi)$ is a coupling function here chosen to be $A(\phi) = e^{k\phi/m_p} = e^{\phi/M}$. The potential $V(\phi)$ needs to be a runaway potential. Originally, a Ratra–Peebles potential parametrized by 2 constants α and Λ was considered $V(\phi) = \frac{\Lambda^{4+\alpha}}{\phi^\alpha}$ (see Ratra and Peebles 1988). It can be shown that the evolution of the scalar field is driven by an effective potential that depends on the local matter density, which is the seed of the screening mechanism.

Since this class of theory is built in order to make a link between local scales and large scales, it is interesting to combine local measurements with cosmological observations. For example, in Hees and Füzfa (2012), Supernovae Ia observations are combined with the γ PPN measurement in order to constrain the parameters of this theory. Figure 3 shows the constraints obtained using the two types of observations. The cosmological constraint is nearly insensitive to the value of the coupling parameter k and is therefore very similar to the constraint obtained for quintessence fields. On the other hand, the Cassini constraint is very sensitive to the value of the parameter k. As can be seen in this Figure, the original chameleon theory is not able to explain the cosmological observations while satisfying the local tests of gravitation. A further detailed analysis (see Hees and Füzfa 2012) shows that for parameters that can explained the Supernovae Ia observations, the screening mechanism is not at work in the Solar System, therefore producing unacceptable deviations. This data analysis was confirmed by analytical calculations in Wang et al. (2012).

One solution to save the chameleon theory is to consider a more complex potential like e.g. $V(\phi) = \Lambda^4 e^{\Lambda^\alpha/\phi^\alpha} \approx \Lambda^4 + \frac{\Lambda^{4+\alpha}}{\phi^\alpha}$ (see e.g. Zlatev et al. 1999) which from a practical point of view corresponds to adding a cosmological constant to the standard Ratra–Peebles potential. In this case, the first term is responsible for DE while the Ratra–Peebles potential is responsible for the screening mechanism. This model is able to drive the cosmic acceleration observed today if $\Lambda \approx 2.4$ meV (see Hamilton et al. 2015), leaving the k parameter nearly unconstrained. It is interesting to notice that in this case, atom interferometry measurements can provide a very stringent constraint. Indeed, it has been proposed by Burrage et al. (2015) to use atom interferometry in a large vacuum chamber. The idea is to screen the scalar field generated outside the chamber and to look for a chameleon field generated by atoms inside the chamber. Since atoms are very small, these chameleon fields will not be screened and

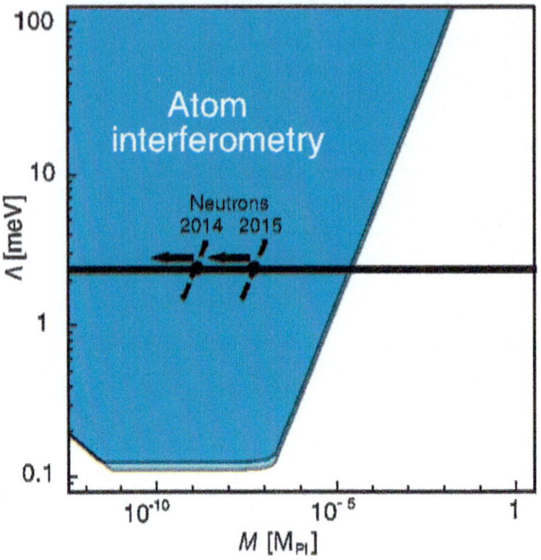

Fig. 4 95% confidence region of exclusion of the chameleon parameters from atom interferometry for $\alpha = 1$. The *light blue area* corresponds to another modeling of the vacuum chamber showing the robustness of the results. The *horizontal thick line* represents the value $\Lambda = 2.4$ meV where the chameleon field would be able to drive the cosmic acceleration. From Hamilton et al. (2015)

can therefore be detected. This measurement has been performed in Berkeley (see Hamilton et al. 2015) and excludes an impressive part of the parameter space as can be seen from Fig. 4.

In conclusion, the chameleon field is also an interesting theory where the combination of data from local scales and larger scales is highly fruitful. This theory is not ruled out per se but the open parameter space becomes very narrow. The question of fine-tuning can therefore be raised.

2.3.3 Modified Newtonian Dynamics (MOND)

The last example of a theory that benefits from a combined analysis of local scale measurements with larger scales observations is the MOND phenomenology. As already mentioned in Sect. 2.2 above, this phenomenology depends on an interpolating function ν making the transition between the Newtonian gravity and the MONDian gravity and on a MOND acceleration scale a_0. Different functions ν have been considered in the literature (see e.g. Famaey and McGaugh 2012) in order to explain galactic observations (like e.g. rotation curves). As developed in Sect. 2.2 above, it has been recently shown that this phenomenology produces an effect in the Solar System: the External Field Effect. This effect is parametrized by a quadrupolar moment Q_2 that depends on the interpolating function ν and on the acceleration scale a_0. This quadrupolar moment has been constrained using Cassini radioscience tracking measurements (see Hees et al. 2014b and Eq. (16)). Therefore, this theory will also benefit from a combined analysis including Solar System measurements and galactic rotation curve observations. This analysis is presented in Hees et al. (2016b). The idea is to use observations of 27 galactic rotation curves (using data from Swaters et al. 2010) and to fit them with different MOND phenomenology. In this analysis, 31 different interpolating functions ν have been considered. For each of these interpolating functions, an optimal value of the MOND acceleration a_0 scale has been determined. This value leads to a prediction for the MOND quadrupolar moment Q_2 that arises in the Solar System. It is then possible to compare this value with the Cassini constraint (see Eq. (10)). Out of the 31 functions considered, only 10 (if one choses a very conservative approach) are able to explain the rotation curves and the Solar System observations. It is worth mentioning that the inclusion of the last years of Cassini tracking measurements (the analysis from Hees et al. 2014c includes observations up to 2013 only) further improves the constraint on Q_2 (W. Folkner 2016, private communication) and will reduce the number of viable MOND interpolating functions. This shows once again the interest of combining observations from different scales.

2.3.4 Summary

In this section, we have presented a few examples where the combination of measurement at local scales (in the lab or at the scale of the Solar System) with larger scales observations (either galactic observations or cosmological observations) leads to very fruitful results. Indeed, the modifications of the gravitation theory made in order to explain galactic or cosmological observations may produce detectable effects at smaller scales (sometimes unexpected). It is therefore highly interesting to combine all these observations in order to facilitate a detection of a deviation from GR or to produce the most stringent constraints within the zoo of alternative theories of gravitation. The examples sketched in this section show the power of such combined analysis and gives strong motivation to pursue these.

2.4 Conclusion

In this section we have provided some of the theoretical background that provides the motivation for experimental tests of gravitation, and is generally used to evaluate and compare such tests. Since the advent of general relativity, gravitation is intimately related to space–time, and the most accurate space–time metrology is based on atomic clocks and the exchange of electromagnetic signals. It is thus no surprise that many of the best tests of gravitation are based on high accuracy atomic clocks on ground and in space. This will be further illustrated by some specific examples in the next section.

3 Some Past, Present and Near Future Experiments

In this section we describe four experiments in fundamental physics that rely on clocks and timing measurements. Three of them concern a test of the Einstein Equivalence Principle via the gravitational redshift (Local Position Invariance) as described in Sect. 2.1. The last one is a test of the Einstein field equations in the parametrized post Newtonian (PPN) framework described in Sect. 2.2. Those experiments have either provided the most stringent constraints in their respective domain to date, or are expected to do so in the very near future (next few years), and are in that sense key experiments in the domain of fundamental physics.

3.1 Gravity Probe A (GP-A)

The most precise test of the gravitational redshift to date has been realized with the Vessot–Levine rocket experiment in 1976, also named the Gravity Probe A (GP-A) (Vessot and Levine 1979; Vessot et al. 1980, 1989). The goal of this experiment was to perform a gravitational redshift test by monitoring the variation of an onboard hydrogen-maser (H-maser) when compared to ground H-masers.

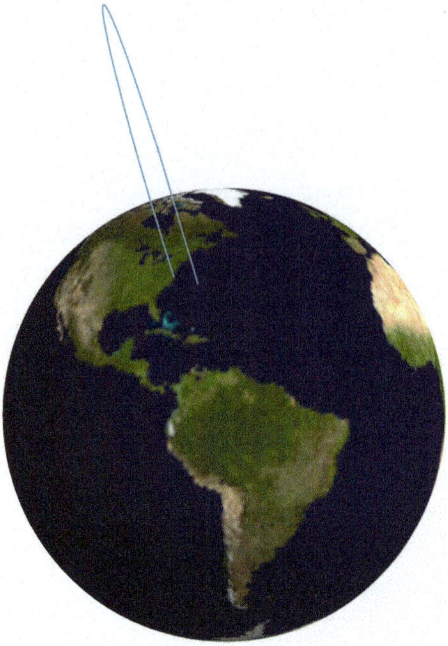

Fig. 5 Approximate trajectory of GP-A

3.1.1 Description of the Experiment

On June 18, 1976, a Scout D rocket was launched on a parabolic trajectory from the Wallops island, Virginia, carrying an atomic H-maser clock as the payload. The frequency of signals from the onboard clock was monitored by two H-masers on the ground at Merrit Island, Florida. The first-order Doppler shift in the signals was eliminated by a two-way transponder link to the payload. We show on Fig. 5 the approximate trajectory of GP-A in inertial space. It was launched at 11:41 GMT and the impact was at 13:37 GMT at latitude 29°44 N, longitude 47°75 W. The experiment was therefore completed after a total time of 1 h 58 min. However the usable data spans over around 1 h 20 min.

The main parts of the payload consisted of a H-maser clock, an S-band transmitter phase-locked to the maser, and an S-band transponder (see Fig. 6). The payload weighed 57 kg and was spin stabilized to maintain its orientation in inertial space. The S-band signals were received and transmitted from a common antenna located on the spin axis at the aft end of the payload. The principle of the Doppler-cancelling system is shown on Fig. 6: the uplink signal 1 is phase-locked to the ground H-maser, sent to the onboard transponder and sent back to the ground, which gives a beat frequency $(f_0)'' - f_0$. The downlink signal 3, phase-locked to the onboard H-maser, is mixed with the ground H-maser frequency and gives a beat frequency $(f_0)' - f_0$. Half of the beat frequency $(f_0)'' - f_0$ is subtracted from the beat frequency $(f_0)' - f_0$, thus cancelling the first-order Doppler effect. Moreover, the frequencies of the three signals are chosen such that the first-order ionospheric delay is canceled. The remaining error coming from the ionosphere is below 5×10^{-15} in relative frequency.

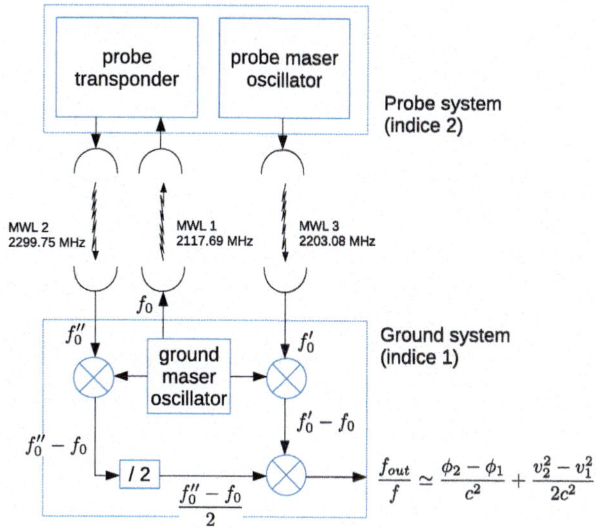

Fig. 6 Schematic concept of the Doppler-cancelling system

3.1.2 Predicted Frequency Shift

The proper time to coordinate time ratio of an ideal clock can be written in a quasi-inertial coordinate system centered on Earth as:

$$y \equiv \frac{d\tau}{dt} = 1 - \frac{GM}{rc^2} - \frac{v^2}{2c^2} \tag{17}$$

where GM is the gravitational parameter of the Earth, c is the velocity of light, r is the distance of the clock from the Earth center, and v is the coordinate velocity of the clock. The two small terms at the r.h.s. of the equation are usually called gravitational redshift and second order Doppler effect.

We show on Fig. 7 the predicted frequency shift during the GP-A experiment based on an approximated trajectory. We can see that the variation of the second order Doppler effect is almost the same as the gravitational redshift. This is generally the case for an almost Keplerian orbit, because the reduced energy $E = v^2/2 - GM/r$ is conserved. Therefore we get for an almost Keplerian orbit:

$$\delta\left(\frac{v^2}{2c^2}\right) \simeq \delta\left(\frac{GM}{rc^2}\right) \tag{18}$$

where $\delta()$ is a small variation with respect to time. This is the case for GP-A because the trajectory is quasi-Keplerian. It is almost a parabola as it has a very high eccentricity ($\simeq 0.999$).

3.1.3 Error Budget

We reproduce the error budget of GP-A from Vessot et al. (1989) Table 1. The combined random and systematic errors sum up to 15.5×10^{-15}, whereas the variation of gravitational redshift over useful data is estimated to be 1.1×10^{-10}, which can be seen on Fig. 7. Therefore the estimated accuracy on the LPI violation is 1.4×10^{-4}.

The limiting error comes from the overall microwave system noise and maser instability. That is why Vessot et al. (1989) proposed to perform a similar experiment in orbit. Averaging

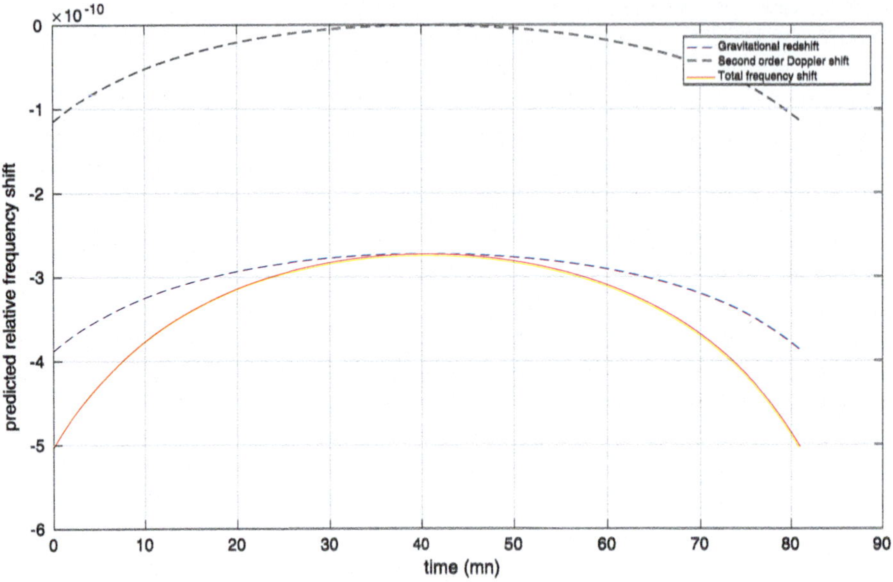

Fig. 7 Predicted relative frequency shift for the GP-A experiment, based on an approximated trajectory

Table 1 Error budget of the GP-A experiment assuming a total time of around 2 hours, and of a similar experiment in orbit assuming a 18–24 hours orbital period (reproduced from Vessot et al. 1989)

	GP-A (2 hours)	Orbital test (18–24 hours)
Random noise		
Overall microwave system noise including maser instability	13.0×10^{-15}	$\lesssim 10^{-16}$
Uncancelled atmospheric noise	2.5×10^{-15}	2.5×10^{-15}
Uncancelled ionospheric noise	5.0×10^{-15}	0.2×10^{-15}
Systematic errors		
Ground clock	3.0×10^{-15}	1.0×10^{-15}
Space clock	4.5×10^{-15}	1.0×10^{-15}
Errors coming from position and velocity uncertainties	3.3×10^{-15}	1.0×10^{-15}
Combined random and systematic errors	15.5×10^{-15}	4.7×10^{-15}
Variation of gravitational redshift over useful data	1.1×10^{-10}	4.8×10^{-10}
Predicted error in a single redshift measurement	141×10^{-6}	9.9×10^{-6}

over a longer time, the system noise is below 10^{-16}. Vessot evaluated the random noise to be 3.2×10^{-15} for one orbit. Averaging over 30 orbits, the random error becomes $3.2 \times 10^{-15}/\sqrt{30} \simeq 5.8 \times 10^{-16}$. Assuming a contribution of systematic errors of 1.7×10^{-15} and a variation of gravitational redshift of 4.8×10^{-10}, one could get a relative error of 3.8×10^{-6} on the violation of LPI. We will see in the next section that we can perform such a test with the eccentric Galileo satellites. However the errors coming from position and velocity uncertainties will be higher because there is no two-way transfer, so they impact directly the determination of the onboard clock frequency.

3.2 Galileo Satellites Doresa and Milena

Galileo satellites Doresa and Milena (or 5 & 6) satellites were launched on August, 30th 2014. Because of a technical problem, the launcher brought them on a wrong, elliptic orbit. An elliptic orbit induces a periodic modulation of the gravitational redshift (Ashby 2003). With the good stability of recent GNSS clocks one can then test this periodic modulation to a very good level of accuracy. The Galileo 5 and 6 satellites, with their large eccentricity and on-board H-maser clocks, are hence perfect candidates to perform this test. Contrary to the GP-A experiment, it is possible to integrate the signal on a long duration, therefore improving the statistics. Data collection is ongoing and we expect first result of this experiment in 2016/2017. In the following, we assess the sensitivity of the gravitational redshift test that can be performed with Galileo satellites 5 and 6 using numerical simulations. This is based on article (Delva et al. 2015).

3.2.1 Description of the Experiment

The experiment proposed in Delva et al. (2015) requires an accurate knowledge of the frequency of the satellite clock as it orbits the Earth. These data are made available by several analysis centers (ACs) of the International GNSS Service, IGS (Dow et al. 2009) in the framework of the Multi-GNSS-Experiment, MGEX (Montenbruck et al. 2013). MGEX applies the standard IGS processing chain to new GNSS (Galileo, Beidu) and regional services (e.g., QZSS for Japan). All active Galileo satellite orbits, clock corrections, and inter-system biases are then available for public use through the MGEX product directory at the IGS data center CDDIS, which is updated on a regular basis.

Since each AC is following a different strategy, as an example we shall focus on the processing performed at the CODE (Center for Orbit Determination in Europe) analysis center (Dach et al. 2013). In this particular case, the whole solution is based on a two steps process. First, satellite orbits are determined using a double difference network solution. This technique allows to eliminate or reduce several biases by combining observations among a pair of receivers and a pair of satellites. All clock errors are eliminated in the combination while the remaining parameters (including ephemerides, atmospheric and antenna parameters, etc.) are estimated for all stations and satellites involved. The latter are then used as a framework for the clock solution, which is based on a zero-difference processing. For more general information about the IGS and its products and data centers, the reader can refer to Dow et al. (2009).

For the test proposed in Delva et al. (2015), the orbit solution of Galileo 5 and 6 satellites would be used to calculate the behavior of the on-board clocks and the gravitational redshift as predicted by GR (Will 1993, 2014). The latter would then be compared to the clock solution from the IGS processing to recover any violation of the LPI. Several conditions need to be satisfied by the data to assure the success of the experiment. The availability of continuous orbit and clock parameters is paramount and is assured by the IGS guidelines for the clock and orbit products. Orbits and station coordinates should be unaffected by a putative violation of the gravitational redshift. This is assured as a result of the processing outlined above, since both are derived by a double-difference solution which eliminates all clock parameters. Finally, the reference clock for the solution is always chosen among ground station H-maser clocks (Bock et al. 2009), so that it is unaffected by a possible LPI violation. Below we study the attainable sensitivity of the test using simulated data.

Table 2 Orbital parameters of satellites Galileo 5 and 6 (ESA, personal communication). Their orbits were circularized so that they can be used for positioning. Both orbits are now very similar

Name	e	a (km)	i (°)
Galileo 5 & 6 final orbit	0.1561	27977	49.7212
Galileo 5 & 6 initial orbit	0.2330	26192	49.7740

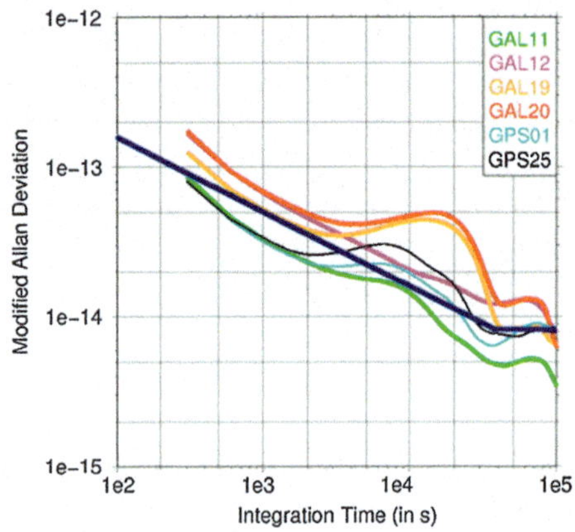

Fig. 8 MDEV for Galileo In-Orbit Validation (IOV) (GAL11-12-19-20) and for GPS Block II-F (PRN GPS01-25) from Prange et al. (2013). Superimposed in *blue* is the MDEV of the simulated clock noise

3.2.2 Simulation: Orbits, Clocks and Signal

Orbital parameters of Galileo satellites 5 and 6 used for the simulation of the present work are summarized in Table 2. Their orbits have been circularized, from initial eccentricity $e = 0.2330$ to $e = 0.1561$. However, we will consider both the initial and the final orbits, in order to compare the sensitivity of the redshift test at different eccentricities. Orbits are calculated by solving the Kepler equation for a duration of two years, thus ensuring the numerical stability of the solution with time.

We use the clock solutions provided by CODE to simulate a realistic random noise for Galileo satellite clocks. Figure 8 shows the Modified Allan Deviations (MDEVs) for several GPS and Galileo clocks (Prange et al. 2013) around Day Of Year (DOY) 100/2013, as well as the MDEV of the simulated clock noise (in blue). For an integration time of 1000 s, the MDEVs are between 3×10^{-14} and 7×10^{-14}, and the MDEV of the simulated clock noise is around 5×10^{-14}. The MDEVs decrease like $\tau^{-1/2}$, which is characteristic of white frequency noise. There is a bump at about 21000 s due to systematic effects. Moreover, we add a flicker noise to the simulated clock noise at 8×10^{-15} in the MDEV (flat part of the blue curve on Fig. 8. Although Fig. 8 shows no sign of an increase of noise at large averaging times (indicative of random walk frequency noise or frequency drift), on ground measurements of Galileo H-maser clocks have shown such behavior (Rochat et al. 2012). The latter is anyway not exceeding 5×10^{-15} in MDEV for measurements up to 10^5 s averaging time. At the frequency of our expected periodic signal (2.1×10^{-5} Hz), the corresponding noise is about an order of magnitude below our noise model, and we will therefore neglect it in our analysis.

The ideal signal y is the gravitational part of the relative frequency difference between a ground clock g and the satellite clock s:

$$y = \frac{GM}{c^2}\left(\frac{1}{r_g} - \frac{1}{r_s}\right), \tag{19}$$

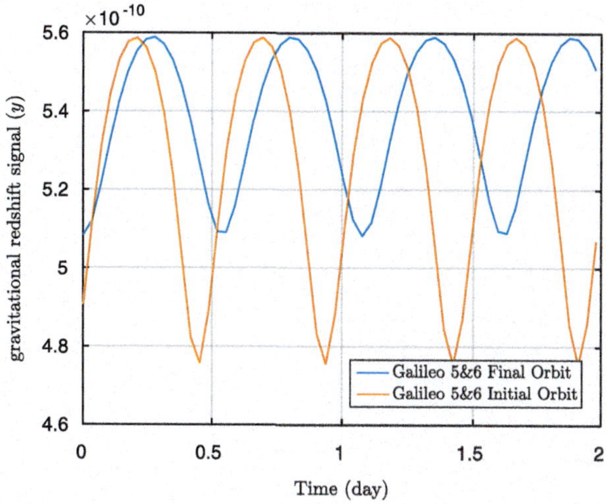

Fig. 9 Gravitational redshift signal with $\alpha = 0$

r_g and r_s are respectively the norm of the position vectors of the ground clock and of the on-board clock. We neglect the noise coming from the ground clock, and we are interested only in the variable part of y. Therefore one can remove the first term in (19), which is constant. We use a simple phenomenological model for a possible violation of the gravitational redshift, characterized by a parameter α which is zero in general relativity (Will 1993, 2014):

$$\tilde{y}(\alpha) = -(1+\alpha)\frac{GM}{c^2 r_s}. \tag{20}$$

The GP-A limit corresponds to $\sigma_\alpha = 1.4 \times 10^{-4}$ (Vessot et al. 1989), where σ_α is the uncertainty on the estimation of the parameter α.

The simulated signal y is shown in Fig. 9. Its amplitude is around twice as large for Galileo satellites 5 & 6 initial orbits than for Galileo 5 & 6 final orbits. The asymmetry between the upper part and the bottom part of the curves is characteristic of the elliptical orbit.

3.2.3 Estimation of the Statistical Sensitivity of the Test

We study the sensitivity of the gravitational redshift test performed with Galileo 5 and 6 through two very different methods: the Fast Fourier Transform (FFT) method and the Linear Least-Square (LSQ) method.

In the FFT method the sensitivity of the gravitational redshift test is calculated as the inverse of the Signal-to-Noise Ratio (SNR) ρ. The SNR can be maximized with matched filtering, i.e. by using a filter optimized for the searched signal and the considered noise. Then the SNR is given (Gardner 1990):

$$\rho^2 = \int_{-\infty}^{+\infty} \frac{|F[y](f)|^2}{S_\epsilon(f)} df, \tag{21}$$

where y and ϵ are the gravitational redshift signal and the clock noise, respectively, $F[y]$ is the Fourier transform of y, and S_ϵ is the Power Spectral Density (PSD) of the clock noise. We implemented a numerical version of this method with the following steps: first we generate

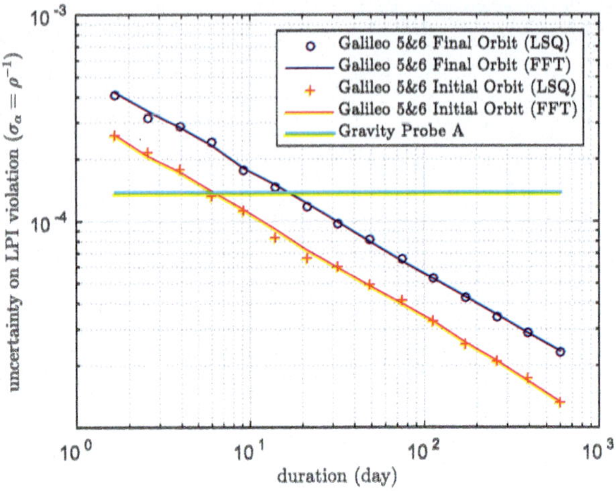

Fig. 10 Statistical sensitivity of the gravitational redshift test with respect to the duration of the experiment. We show the results for the two methods FFT and LSQ, and for the initial and final orbits of Galileo satellites 5 & 6. In *green* is the limit from the Gravity Probe A experiment for reference

100 different sequences of clock noise; we calculate the PSD of each sequence using a fast Fourier transform algorithm and average the 100 different PSD; finally we compute the Signal-to-Noise Ratio (SNR) (21) with the mean PSD of the clock noise, by integrating in the frequency domain $[-F_s/2, F_s/2]$, where F_s is the sampling rate of the signal. This is done for several different durations of the experiment.

The second method implemented consists in using a linear least-square (LSQ) fit combined with a Monte-Carlo method, where we find the minimum of the merit function

$$\chi^2 = \sum_{i=1}^{N} \left[\left(y(t_i) + \epsilon_i \right) - \left(\tilde{y}(\alpha; t_i) + A \right) \right]^2, \tag{22}$$

with respect to α and A. In Eq. (22) N is the number of simulated observables, A is a constant to be estimated and $(\epsilon_i)_{i=1..N}$ is one sequence of simulated clock noise, while $(y(t_i) + \epsilon_i)$ corresponds to the simulated observable at time t_i. The constant A is introduced to remove a constant frequency bias which cannot be measured accurately. Instead of fitting A, it is also possible to remove the mean of $\tilde{y}(\alpha)$ from the model, so that the fitted model has a zero mean, like the noise ϵ. We implemented a numerical version of this method, with the following steps: we generate 100 different sequences of clock noise (we take the same ones as for the FFT method); we estimate α for each clock noise sequence with the linear LSQ method, for different durations of the experiment; finally, for each duration, we calculate the mean of the 100 obtained values of α and their standard deviation σ_α. The standard deviation corresponds approximately to 68% of the obtained values around the mean.

Following Arun et al. (2005) it can be shown that for high SNR and in the case of a one parameter determination, the inverse of the SNR found with matched filtering method is equal to the standard deviation of α found with the LSQ method, i.e. $\sigma_\alpha = \rho^{-1}$. Figure 10 shows the statistical sensitivity of the gravitational redshift test, σ_α, with respect to the duration of the experiment for the two different methods presented above and for the final and initial orbits of Galileo satellites 5 & 6. Also, we verify that the two methods give the same estimate, thus giving us strong confidence in the reliability of our analysis. This result is very promising, as it can be seen that even with the less eccentric orbit, the GP-A limit could be attained in less than a month. However, this result has to be tempered with a study of systematic effects.

3.2.4 Systematic Effects

All systematic effects (i.e. errors that are due to unmodeled deterministic effects) that mimic the gravitational redshift signal may induce a bias in the estimation of α i.e. a "fake" violation of the gravitational redshift. This will ultimately limit the useful duration of noise averaging presented in Fig. 10. The bias on the estimated value of α resulting from unmodeled systematic effects will depend on their magnitude, frequency and phase. We classify systematic effects in four classes: (i) effects acting on the frequency of the reference ground clock; (ii) effects on the links, e.g., mismodeling of ionospheric or tropospheric delays, variations of receiver/antenna delays, multipath effects, etc.; (iii) effects acting directly on the frequency of the space clock, e.g., temperature and/or magnetic field variations on board the Galileo satellites; (iv) orbit modelling errors. The latter directly translate into variations of the measured clock frequencies because the clock solutions are obtained from the one-way signals (so called "zero differences"), hence an orbit error of Δx along the line of sight leads directly to an error in the clock phase estimation of $\simeq \Delta x/c$.

Effects on the ground reference clock are likely to be negligible with respect to the others, and are furthermore likely to be only weakly correlated with the signal we are searching for shown in Fig. 9. They will therefore be neglected in this analysis.

Effects on the link could be significant for certain stations (e.g. under unfavorable multipath environment or atmospheric conditions). However they are expected to vary with elevation and azimuth, and local meteorological conditions, none of which are correlated with the expected signal. Note also that the main signal frequency is 2.1×10^{-5} Hz whereas the 2nd harmonic of any diurnal effect (temperature, humidity etc.) is 2.3×10^{-5} Hz, which indicates that such effects will decorrelate from the signal with as little as 6 days of data. Finally, some significant averaging can be expected given the large number of observing stations.

The environmental sensitivity of the Galileo passive hydrogen masers (PHM) has been characterized on ground (see e.g. Rochat et al. 2012). Temperature sensitivity is $\lesssim 2 \times 10^{-14}$/K and magnetic field sensitivity $\lesssim 3 \times 10^{-13}$/G. Both Galileo 5 and 6 are equipped with on-board temperature sensors, but no in-situ temperature measurements are publicly available. From Fig. 9 we see that the expected peak to peak variation of the redshift signal is about 5×10^{-11}. Given the expected statistical sensitivity of our test (cf. Fig. 10) we want any systematic effect to remain 2×10^{-5} below that level, i.e. $\lesssim 10^{-15}$. This implies that temperature variations at the clock that are correlated with the signal need to be $\lesssim 0.05$ K peak to peak. Whether this is achieved in orbit remains to be seen once temperature measurements are made available. As an example, Montenbruck et al. (2012) mention that for the GPS block IIF "the rubidium clocks of the Block IIF satellites are mounted on a thermally isolated sub-panel (along with the frequency distribution unit) for which a peak-to-peak temperature variation of less than 0.5 K has been observed in SVN62 on-board measurements. In addition, the new clock benefits from a baseplate temperature controller (BTC), which further reduces the sensitivity to external temperature variations by up to a factor of 50." This indicates that peak to peak variations $\lesssim 0.05$ K are realistic. Additionally, significant decorrelation between temperature effects and our signal can be expected, as the former depend on Sun exposure of the satellite and thus have annual spectral components, which are absent in the signal. To evaluate the influence of the magnetic effect on our determination of α we model the Earth magnetic field as a dipole with the field amplitude at the location of the satellite given by

$$|B| = B_0 \left(\frac{R_E}{r} \right)^3 \sqrt{1 + 3\cos^2\vartheta}, \tag{23}$$

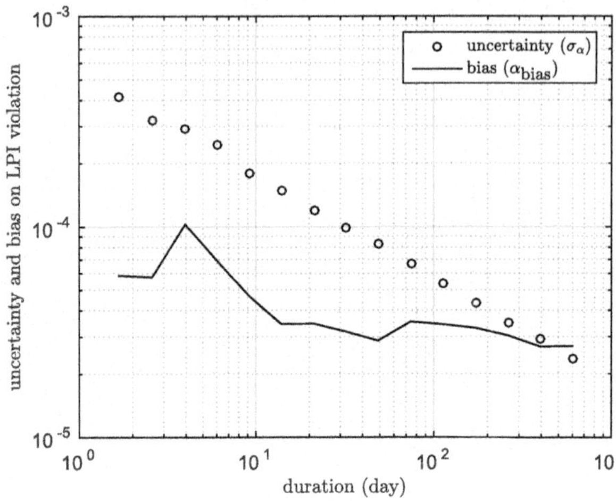

Fig. 11 Statistical uncertainty (*blue circles*) on the LPI violation and bias induced by uncorrected magnetic effect (*black line*) as a function of observation time, for Galileo 5 & 6 final orbits

where $B_0 = 0.312$ G, R_E is the mean radius of the Earth and ϑ is the angle between the position vector of the satellite and the geomagnetic North pole. We then use the clock sensitivity coefficient to add a systematic effect $\epsilon_{sys}(t)$ to our signal, where the time dependence is given by (23) and by the satellite orbit. We determine α from the simulated signal Y using the LSQ method (see above), such that

$$Y(t) = y(t) + \epsilon_{sys}(t) + \epsilon(t), \tag{24}$$

where ϵ is the random clock noise. Figure 11 shows the obtained value of α (the bias on α induced by the systematic effect) as a function of observation time together with the obtained uncertainty from the LSQ-Monte Carlo method. We see that for observation times up to 450 days the magnetic effect remains below the statistical uncertainty. For longer integration times it will become limiting at $\simeq 2.7 \times 10^{-5}$. This is a worst case estimation as it is of course possible to correct at least partly for the magnetic effect and/or simultaneously fit the magnetic effect and α (cf. the discussion of the solar radiation pressure effect below) leading to a better result.

Finally, we look at effects related to orbit modelling errors. The radial part of the resulting orbit error contributes directly to the clock estimation error. Indeed, it has been shown that the clock offset is correlated with satellite laser ranging (SLR) residuals, which can be considered as a measure of the radial orbit error (Montenbruck et al. 2014). The clock error also shows a dependency with the Sun elevation angle, which is the angle between the satellite orbital plane and the direction of the Sun. This clearly indicates systematic effects which are linked to the direction of the Sun, like a mismodeling of Solar Radiation Pressure (SRP) (Montenbruck et al. 2014). Indeed, the use of an enhanced SRP model for orbit determination can divide by a factor of four the effect of SRP mismodeling on the clock estimation. The resulting systematic effect on the clock estimation is evidenced as a bump at half-orbital time scale in the Modified Allan Deviation (MDEV) of Fig. 8. Consequently, we will consider here a class of systematic effects which shows a dependency with the Sun direction, as well as with the mean anomaly of the satellite. For the SRP effect in particular we use a model that reproduces the annual amplitude variation (cf. Fig. 3 Montenbruck et al. 2014) and has a frequency of $n + \omega_a$, where n is the mean motion of the satellite and $\omega_a = 2\pi/\text{year}$:

$$\epsilon_{sys}(t) = A\big(1 + B\big(\cos(\omega_a t + \phi_1) - 1\big)\big)\sin\big((n + \omega_a)t + \phi_2\big), \tag{25}$$

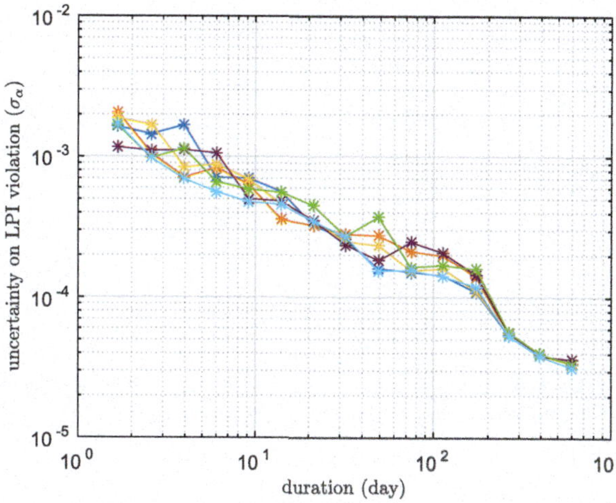

Fig. 12 Statistical uncertainty of the determination of the gravitational redshift in the presence of solar radiation pressure, as a function of observation time, for Galileo 5 & 6 final orbits, and for six different occurrences of the simulated clock noise

with A, B, ϕ_1, ϕ_2 constant. We simulate data with $A = 5 \times 10^{-14}$, $B = 0.45$, $\phi_1 = 0$ that reproduce the results Montenbruck et al. (2014), and different values of ϕ_2. We then use a Bayesian approach[1] with a slice sampling method (Neal 2003) assuming normal distributions and flat priors to simultaneously determine the four parameters of the SRP model (25) and the redshift parameter α, as well as their respective uncertainties. The result is shown in Fig. 12 for six different occurrences of the simulated clock noise. We find that for data spanning more than 1 year, and a "worst case" choice of ϕ_2 the resulting value of α is unbiased with a statistical uncertainty of $\simeq 4 \times 10^{-5}$, slightly increased with respect to the uncertainty when determining α with no SRP effect present in the observable (cf. Fig. 10). After 200 days of integration time, the six occurrences of noise converge to the same values which indicates that the uncertainty is limited by the residual SRP systematic effect. In our estimates we have assumed the "standard" SRP corrections rather than the improved ones discussed Montenbruck et al. (2014). The latter should allow a reduction of the effect by a factor 2–4 (Montenbruck et al. 2014), so our estimates are likely to be again only an upper limit.

3.3 ACES/PHARAO

ACES/PHARAO (Atomic Clock Ensemble in Space/Projet d'Horloge Atomique par Refroidissement d'Atomes en Orbite) is an international space mission of the French and European space agencies (ESA and CNES). It is an international scientific and industrial collaboration with French lead (LKB and SYRTE) that aims at realizing a time scale of high stability and accuracy on board the International Space Station (ISS). The ACES/PHARAO payload will be launched with Space-X and installed onboard the ISS in 2018, with a planned mission duration of 18 months to 3 years.

The on board fractional frequency stability (ADEV) should be better than $\sigma_y = 10^{-13} \times \tau^{-1/2}$, which corresponds to 3×10^{-16} after one day of integration. Time deviation (TDEV) should be better than $\sigma_x = 4.1 \times 10^{-14} \times \tau^{1/2}$ s, which corresponds to 12 ps after one day of integration (see Fig. 13). Absolute frequency accuracy should be around 10^{-16} in fractional

[1] See for example http://fr.mathworks.com/help/stats/examples/bayesian-_analysis-_for-_a-_logistic-_regression-_model.html.

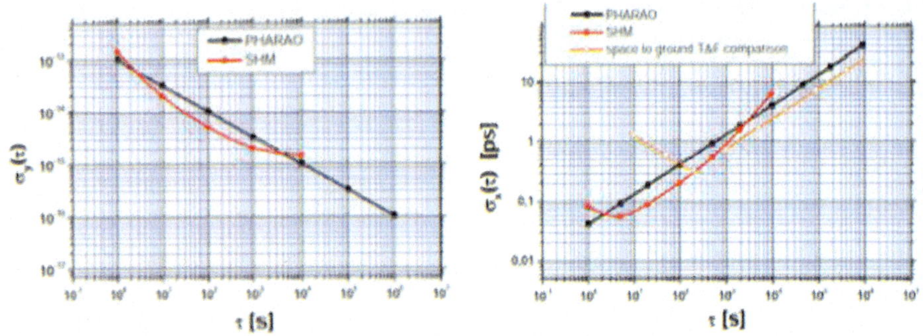

Fig. 13 *Left*: fractional frequency stability (ADEV) of PHARAO and Space Hydrogen Maser (SHM). *Right*: Time stability (TDEV) of PHARAO, SHM and the space-ground time/frequency comparison link

frequency. To achieve these performances, its payload includes the first cold atom clock in space, PHARAO, which is a Cs clock, and a H-maser. Another key element is the microwave link (MWL) that uses radio (Ku and S band) in a two-way configuration to compare space to ground clocks at unprecedented stability and accuracy.

The scientific objectives are:

- Demonstrate the high performance of the atomic clocks ensemble in the space environment and the ability to achieve high stability space-ground time and frequency transfer.
- Perform space to ground clock comparisons with high resolution on a world-wide basis using a link in the microwave domain. The link stability should reach around 0.3 ps after 300 s of integration and around 7 ps after 1 day of integration.
- Perform tests of the Einstein Equivalence Principle (EEP). It will be possible to test Local Lorentz Invariance (LLI) and Local Position Invariance (LPI) to unprecedented accuracy by doing three types of tests: a test of the gravitational redshift, contribute to the search for a variation of fundamental constants, an LLI tests in different theoretical framework (e.g. the SME, see Sect. 2).

Besides these primary objectives, several secondary objectives can be found (Cacciapuoti and Salomon 2009). For example, the measurement of gravitational redshift can be used to measure gravitational potential differences between different clock locations, which is a new type of geodetic measurements using clocks, called chronometric geode (Delva et al. 2014).

LKB and SYRTE are heavily involved in the experiment, from the conception of the cold atom clock to the development and the data analysis. SYRTE is an official ACES Data Analysis Center. Presently we develop a data analysis algorithm in order to process the mission raw data and extract the scientific products of the mission, which are mainly the desynchronization between clocks. In order to test our algorithm, we developed a full simulation of the raw observables, as they will be produced by the MWL modem and instrumentation.

A violation of LPI can be quantified as described in the previous section by a factor α appearing in the standard formula of gravitational redshift. The most stringent limit from previous measurements comes from the 1976 NASA/SAO GP-A rocket experime (Vessot et al. 1989), yielding $|\alpha| \lesssim 1.4 \times 10^{-4}$. This test took advantage of the stability of the spaceborne and ground hydrogen maser clocks, connected by a continuous-wave microwave link, to measure the modulation of the gravitational redshift with altitude. On the contrary, the gravitational redshift in the ACES experiment will only be modulated to around 10%; therefore the high accuracy of the spaceborne (10^{-16}) and ground clocks are used, to measure the absolute redshift which will constrain $|\alpha| \lesssim 3 \times 10^{-6}$, an improvement by a factor 45 compared to the GP-A experiment.

High accuracy comparisons of atomic frequency standards based on different atoms and ions can be interpreted as a test of the variation of fundamental constants. By combining the comparisons between different types of atomic clocks (eg. Rb/Cs, Al+/Hg+, Hg+/Cs, ...), it is possible to disentangle the contributions coming from three different fundamental constants: α_{em} the fine structure constant (electroweak interactions), $\mu = m_e/m_p$ the electron-to-proton mass ratio, and m_q/Λ_{QCD} the quark mass scaled to the quantum chromodynamics mass scale Λ_{QCD} (strong interactions). Present limits can be found (Guéna et al. 2012). The ACES experiment will improve these limits by increasing the number of possible inter-clock comparisons from distant laboratories.

Finally, the ACES experiment will test LLI by testing the independence of the clock synchronization procedure with the orientation of the microwave link. In this experiment, one compares the rest transition frequency of an atom, to its transition frequency when at high velocity as seen by an observer at rest. The expected sensitivity is expected to be comparable to the best present limit obtained from an Ives–Stilwell type experime (Botermann et al. 2014). Both types of tests are complementary.

A test of LLI in the SME can also be done with PHARAO following the test realized on Earth in Wolf et al. (2006a, 2006b). Beyond a faster integration due to the shorter orbital period as already mentioned, the time dependence of the signal will allow a more constraining set of measurements for SME coefficients: the quantization axis of PHARAO being along the ISS trajectory, over one orbital period its direction will be fully reversed twice, giving rise to a comparatively rapid modulation of the expected signal.

3.4 The Cassini Test of Post Newtonian Gravity

On its cruise phase to Saturn the Cassini mission was used to carry out what is still today the most precise test of post-Newtonian gravity in the solar system (Bertotti et al. 2003). It is a test of the Einstein field equations in the parametrized post-Newtonian (PPN) framework that was described in Sect. 2.2. In that framework it sets the most stringent limit on the γ parameter at $\gamma - 1 = (2.1 \pm 2.3) \times 10^{-5}$. Additionally to providing the best constraint in the solar system on that parameter, the Cassini experiment has also been used at galactic and cosmological scales to provide combined constraints on theories at large scales (cf. Sect. 2.3). The experiment measured the travel time of radio signals between the NASA deep space network (DSN) antennas and the Cassini spacecraft as the line of sight approached the limb of the sun. The good performance of the atomic clocks in the NASA deep space network (DSN) were key to its success.

3.4.1 Principle of the Experiment

During its cruise phase to Saturn the Cassini spacecraft passed almost behind the sun (as seen from the Earth) on 21st June 2002. During the hours before and after closest approach the Radio links between the Earth antennas of the NASA deep space network (DSN) and the Cassini spacecraft were used to precisely measure the frequency shift of radio signals sent to the spacecraft and back. In general relativity as well as all metric theories of gravitation, e.g. PPN, and electromagnetic signal experiences a gravitational delay (also known as Shapiro delay) when passing close to a gravitating body. That delay can be measured by monitoring the return travel time of the signal as the line of sight to the reflector passes close to a gravitational source. The derivative of that delay induces a frequency shift $y = \frac{\Delta f}{f}$ in the signal given by

$$y(t) = -4(1+\gamma)\frac{GM_S}{c^3}\frac{1}{b(t)}\frac{db(t)}{dt} \tag{26}$$

where, M_S is the mass of the sun, $b(t)$ is the impact parameter (point of closest approach) of the light signal and $\gamma = 1$ in GR. The factor 2 difference with respect to Eq. (2) of Bertotti et al. (2003) is a misprint in Bertotti et al. (2003), which however has no further consequences in the analysis there. The point of closest approach occurred with a minimal impact parameter of 1.6 solar radii.

3.4.2 Results

The DSN data was fitted with a 12 parameter model that includes γ, the six initial conditions of the spacecraft trajectory, three parameters for the non-gravitational acceleration due to thermal emissions, and two parameters characterizing the reflectivity of the spacecraft antenna (to correct for solar radiation pressure). The main noise sources were measurement noise including that of the DSN clocks. Systematic effects from dispersion (solar corona and Earth's ionosphere) and troposphere were removed from the data using a dispersion free (to first order) combination of Ka and X band signals, and modeling of the troposphere from meteorological data taken at the DSN sites. No estimate on the residuals dispersion and tropospheric effects, that could mask a putative violation of GR, are available in Bertotti et al. (2003). The other main source of systematic errors is the non-gravitational acceleration of the spacecraft due to thermal radiation and solar pressure. This was accounted for by fitting appropriate parameters in the global model. Additionally, in order to exclude possible effects from mismodelling, different models were used yielding all similar results. The final result of Bertotti et al. (2003) is then

$$\gamma - 1 = (2.1 \pm 2.3) \times 10^{-5} \tag{27}$$

With the uncertainty entirely due to statistics and limited by measurement noise, which includes the noise of clocks in the DSN stations.

4 Conclusion and Outlook

We conclude this paper by describing briefly some space projects in the more distant future that will probe fundamental physics using high accuracy clocks on ground and in space.

The Space Optical Clock (SOC) proposal aims at putting an optical lattice clock onboard the International Space Station (ISS) in a 2020–2025 time frame (Bongs et al. 2015; Origlia et al. 2016). Additionally the mission is planned to carry an improved version of the Microwave link (MWL) of ACES as well as an optical link for high performance space-ground clock comparisons. The objectives are similar to the ACES mission objectives (see above) but with an improvement in performance by about a factor 10. That improvement is expected to come from the accuracy of the clock (at the 10^{-17} level in fractional frequency) and the better performance of the time/frequency transfer links. Currently the project is undergoing a definition and assessment (phase A) study at ESA, expected to last until late 2017.

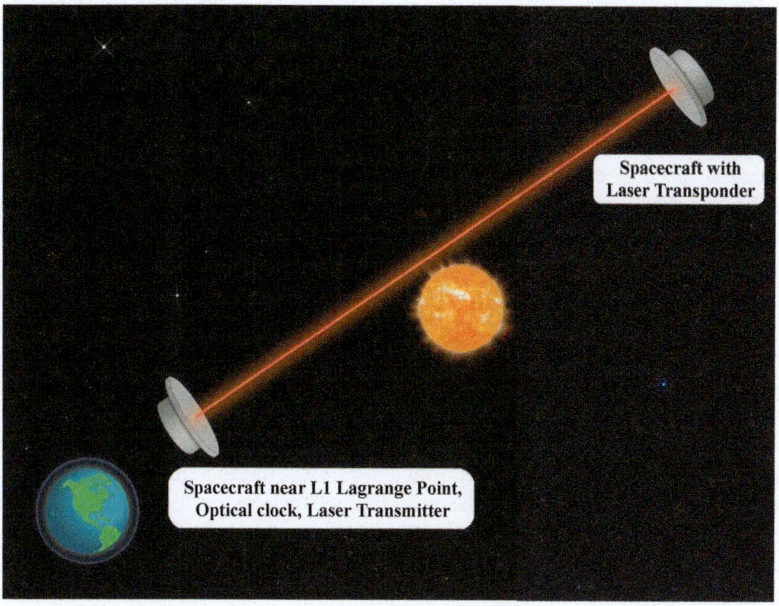

Fig. 14 Principle of the gravitational time delay experiment

The gravitational time delay experiment (Ashby et al. 2010) proposes to measure the gravitational time delay of an electromagnetic signal to unprecedented accuracy. When interpreted in the PPN framework (see Sect. 2.2) it will set a limit on the PPN parameter γ of $\gamma - 1 \leq 10^{-8}$. This corresponds to an improvement on the state of the art (see above) by over three orders of magnitude. The principle of the mission is shown on Fig. 14. A laser link between two spacecraft on either side of the sun is used to measure the return travel time of light signals. One of the spacecraft carries a highly stable optical clock, the other one a laser transponder that returns the signal. Both spacecraft are drag-free controlled thereby avoiding effects coming from non-gravitational accelerations.

Extensive simulations were carried out to study the performance of such a mission. The baseline of those simulations was a 5 year mission, with three occultation events 1, 3 and 5 years after launch. The assumed clock stability of $1 \times 10^{-15}/\sqrt{\tau}$ (Allan deviation), which is already surpassed by best ground clocks. The drag free performance assumed in the simulations was 10^{-13} m/s$^2/\sqrt{\text{Hz}}$, which is an order of magnitude less demanding than the recently demonstrated performance of the LISA-Pathfinder spacecraft (Armano et al. 2016). Finally the measurement noise of the laser return-time measurement was assumed at 0.02 ps$/\sqrt{\text{Hz}}$, which is more than six orders of magnitude less demanding than e.g. the laser link of the LISA mission.

Although similar in principle to the Cassini experiment described in the previous section, a number of modifications will allow significant improvements. Apart from the lower noise levels of the measurement systems, the absence of the atmosphere, drag-free spacecraft and optical frequency of the signal (rather than radio) should allow better control of the main systematics. In particular the optical frequency will strongly decreases the effect of the solar corona and will therefore allow a minimization of the impact parameter, which increases the signal.

In summary, the mission is expected to allow a test of the Einstein field equations at unprecedented levels. In the PPN framework it will lead to an improvement on the knowledge of the γ parameter by over three orders of magnitude. It requires only state of the art performance of present day technology. The main challenge will be the making that technology, and in particular the optical clock, sufficiently small and robust for space operation.

E-GRIP (Einstein Gravitational RedshIft Probe) is a study case of a fast, relatively cheap and high TRL mission to test the Einstein Equivalence Principle (EEP) via the gravitational redshift of a space clock in a highly eccentric Earth orbit. The principle is very similar to the test using the eccentric Galileo 5 & 6 satellites (see Sect. 3.2). It aims at using well developed technology: the space hydrogen maser (SHM), Microwave link (MWL), and laser timing system (ELT) of the ACES mission (see Sect. 3.3), adapted for the different orbit where necessary. The main gain comes from the highly elliptic orbit. Two scenarios were considered, with semi-major axes equal to 24450 km and 35000 km, and eccentricities of 0.636 and 0.800. The goal is a test of the gravitational redshift with a relative uncertainty of 4×10^{-7}, which represents an improvement by about a factor 350 with respect to the best present knowledge from the Gravity Probe-A experiment (see Sect. 3.1), and by about a factor 5 with respect to the expected result from ACES (Sect. 3.3). Although the latter is only a modest gain, it can be obtained from a high TRL, cheap and rapid mission.

The Space–Time Explorer and QUantum Equivalence Space Test (STE–QUEST) space mission (Altschul et al. 2015; STE-QUEST 2013) is an M class mission proposal that was pre-selected by the European Space Agency in 2010 together with four other missions for the cosmic vision M3 launch opportunity. It carries out tests of different aspects of the Einstein Equivalence Principle using atomic clocks, matter wave interferometry and long distance time/frequency links, providing fascinating science at the interface between quantum mechanics and gravitation that cannot be achieved, at that level of precision, in ground experiments. Since 2010 STE-QUEST has been studied in different configurations and mission scenarios. The most extensive version includes two main payload elements: A two species (^{87}Rb–^{41}K, or ^{87}Rb–^{85}Rb) atom interferometer that carries out a test of the universality of free fall in the quantum regime (Bose–Einstein Condensates in large quantum superpositions) and an evolved version of the PHARAO cold-atom space clock from ACES (Sect. 3.3) for tests of the gravitational redshift. Additionally it will carry an improved version of the ACES MWL and a new type of optical link for high performance ground-space time/frequency comparisons. The considered mission scenarios have highly elliptical orbits (700×51000 km in the M3 version, 2500×33600 km in the M4 version) which allow large gravitational field variations, long common view times for ground clocks when the satellite is at apogee, and strong gravitational field at perigee for the UFF test. STE-QUEST will test UFF at a level of 10^{-15}, two orders of magnitude better than the best present result using macroscopic test-objects, and at least 6 orders better than present UFF tests in the quantum regime. It will also improve bets present results on redshift tests (in the field of the Earth, Sun, and Moon) by up to 4 orders of magnitude. Additionally tests of Lorentz invariance in the SME framework are planned that will improve knowledge of several SME coefficients by up to 4 orders of magnitude. Thus STE-QUEST is a mission that tests all aspects of the Einstein Equivalence Principle, searching for any deviation at sensitivities well beyond what is achieved today.

In summary, fundamental physics in space is a highly active field of research that thrives on technological developments, in particular of clocks, long distance radio and optical links, and inertial sensors (accelerometers, gyroscopes). It is impossible to describe all missions and projects in the field, and in this paper we have mentioned only some of the most important ones, and in particular those that rely strongly on clocks and time/frequency transfer methods to achieve their objectives. Very recently (2015/2016) there have been two missions launched that are critical to the field: LISA-Pathfinder that demonstrated unprecedented levels of drag-free technology and performance (Armano et al. 2016), and MICROSCOPE that will carry out the first space-test of the Universality of Free Fall (see Sect. 2.1) improving the best present measurements by about two orders of magnitude. This shows the vitality of the field and many more exciting results are expected in the near and more distant future, which will hopefully pave the way towards the new physics beyond general relativity and the standard model of particle physics. And it is likely that high performance clocks will continue to play a major role in that process.

References

E. Adelberger et al., Annu. Rev. Nucl. Part. Sci. **53**, 77 (2003)
E. Adelberger et al., Prog. Part. Nucl. Phys. **62**, 102 (2009)
S. Aghion et al., J. Instrum. **8**, 8013P (2013)
J.-M. Alimi, A. Füzfa, Int. J. Mod. Phys. D **16**, 2587 (2007)
J.-M. Alimi, A. Füzfa, J. Cosmol. Astropart. Phys. **9**, 14 (2008)
B. Altschul et al., Adv. Space Res. **55**, 501–524 (2015)
M. Armano et al., Phys. Rev. Lett. **116**, 231101 (2016)
K.G. Arun et al., Phys. Rev. D **71**(8), 084008 (2005)
A. Arvanitaki et al., Phys. Rev. D **91**, 015015 (2015)
A. Arvanitaki et al., Phys. Rev. Lett. **116**, 031102 (2016)
N. Ashby, Living Rev. Relativ. **6**, 1 (2003)
N. Ashby et al., Phys. Rev. D **75**, 022001 (2007a)
N. Ashby et al., Phys. Rev. Lett. **98**, 070802 (2007b)
N. Ashby et al., in *Relativity in Fundamental Astronomy, Proceedings IAU Symposium*, vol. 261, 2009, ed. by S.A. Klioner, P.K. Seidelman, M.H. Soffel (2010), p. 414
P. Avelino et al., Phys. Rev. D **64**, 103505 (2001)
A. Avilez-Lopez et al., J. Cosmol. Astropart. Phys. **6**, 044 (2015)
Y. Bai et al., J. Cosmol. Astropart. Phys. **10**, 029 (2015)
Q. Bailey, A. Kostelecky, Phys. Rev. D **74**, 045001 (2006)
Q. Bailey et al., Phys. Rev. D **88**, 102001 (2013)
J. Battat et al., Phys. Rev. Lett. **99**, 241103 (2007)
J. Bekenstein, Phys. Rev. D **25**, 1527 (1982)
J. Bekenstein, J. Magueijo, Phys. Rev. D **73**, 103513 (2006)
B. Bertotti et al., Nature **425**, 374 (2003)
L. Blanchet, J. Novak, Mon. Not. R. Astron. Soc. **412**, 2530 (2011)
H. Bock et al., J. Geod. **83**, 1083–1094 (2009)
K. Bongs et al., C. R. Phys. **16**, 553–564 (2015). arXiv:1503.08457
B. Botermann et al., Phys. Rev. Lett. **113**(12), 120405 (2014)
A. Bourgoin et al., arXiv:1607.00294 (2016)
C. Brans, R. Dicke, Phys. Rev. **124**, 925 (1961)
A. Brillet, J. Hall, Phys. Rev. Lett. **42**, 549 (1979)
J.P. Bruneton, G. Esposito-Farèse, Phys. Rev. D **76**, 124012 (2007)
C. Burrage et al., J. Cosmol. Astropart. Phys. **3**, 042 (2015)
L. Cacciapuoti, C. Salomon, Eur. Phys. J. Spec. Top. **172**, 57 (2009)
S. Carroll, Phys. Rev. Lett. **81**, 153067 (1998)
S. Carroll et al., Phys. Rev. Lett. **103**, 011301 (2009)
S. Carroll et al., Phys. Rev. D **81**, 063507 (2010)
T. Clifton et al., Phys. Rep. **513**, 1 (2012)
D. Colladay, A. Kostelecky, Phys. Rev. D **55**, 6760 (1997)
D. Colladay, A. Kostelecky, Phys. Rev. D **58**, 116002 (1998)
R. Dach et al., Center for orbit determination in Europe (CODE). Technical report, International GNSS Service: Technical Report 2012, ed. by R. Dach, Y. Jean (AIUB), IGS Central Bureau (2013)
T. Damour, Class. Quantum Gravity **29**, 184001 (2012)
T. Damour, J. Donoghue, Phys. Rev. D **82**, 084033 (2010)
T. Damour, A. Polyakov, Nucl. Phys. B **423**, 532 (1994)
T. Damour et al., Phys. Rev. Lett. **64**, 123 (1990)
P. Delva et al., in *Frontiers in Relativistic Celestial Mechanics. Volume 2: Applications and Experiments*, ed. by S.M. Kopeikin (de Gruyter, Berlin, 2014)
P. Delva et al., Class. Quantum Gravity **32**(23), 232003 (2015)

A. Derevianko, M. Pospelov, Nat. Phys. **10**, 933 (2014)
J.M. Dow et al., J. Geod. **83**, 191–198 (2009)
A. Einstein, Ann. Phys. **340**, 898 (1911)
A. Einstein, Ann. Phys. **154**, 769 (1916)
B. Famaey, S. McGaugh, Living Rev. Relativ. **15**, 10 (2012)
A. Fienga et al., Celest. Mech. Dyn. Astron. **123**, 325 (2015)
E. Fischbach, C. Talmadge, *The Search for Non-Newtonian Gravity* (Springer, Berlin, 1999)
J. Frieman, B. Gradwohl, Phys. Rev. Lett. **67**, 2926 (1991)
A. Füzfa, J.-M. Alimi, Phys. Rev. D **75**, 123007 (2007)
W.A. Gardner, *Introduction to Random Processes: With Applications to Signals and Systems*, vol. 31 (McGraw–Hill, New York, 1990)
R. Godun et al., Phys. Rev. Lett. **113**, 210801 (2014)
B. Gradwohl, J. Frieman, Astrophys. J. **398**, 407 (1992)
J. Guéna et al., Phys. Rev. Lett. **109**, 080801 (2012)
P. Hamilton et al., Science **349**, 849 (2015)
A. Hees et al., arXiv:1604.08514 (2016a)
A. Hees, A. Füzfa, Phys. Rev. D **85**, 103005 (2012)
A. Hees, A. Füzfa, in *Thirteenth Marcel Grossmann Meeting: On Recent Developments in Theoretical and Experimental General Relativity, Astrophysics and Relativistic Field Theories* (2015), p. 1140
A. Hees, O. Minazzoli, arXiv:1512.05233 (2015)
A. Hees et al., Class. Quantum Gravity **29**, 235027 (2012)
A. Hees et al., in *Proceedings of the Journées 2013 "Systèmes de Référence Spatio-temporels"* (2014a), p. 241. arXiv:1403.1365
A. Hees et al., Phys. Rev. D **89**, 102002 (2014b)
A. Hees et al., Phys. Rev. D **90**, 124064 (2014c)
A. Hees et al., in *Proceedings of the Annual Meeting of the SF2A* (2015a), p. 125
A. Hees et al., Phys. Rev. D **92**, 064049 (2015b)
A. Hees et al., Gen. Relativ. Gravit. **47**, 9 (2015c)
A. Hees et al., Mon. Not. R. Astron. Soc. **455**, 449 (2016b)
M. Hohensee et al., Phys. Rev. Lett. **106**, 151102 (2011)
M. Hohmann, Phys. Rev. D **92**, 064019 (2015)
N. Huntemann et al., Phys. Rev. Lett. **113**, 210802 (2014)
L. Iorio, Class. Quantum Gravity **29**, 175007 (2012)
H. Ives, G. Stilwell, J. Opt. Soc. Am. **28**, 215 (1938)
M.T. Jaekel, S. Reynaud, Class. Quantum Gravity **22**, 2135 (2005)
M.T. Jaekel, S. Reynaud, Class. Quantum Gravity **23**, 777 (2006)
R. Kennedy, E. Thorndike, Phys. Rev. **42**, 400 (1932)
M. Kesden, M. Kamionkowski, Phys. Rev. Lett. **97**, 131303 (2006a)
M. Kesden, M. Kamionkowski, Phys. Rev. D **74**, 083007 (2006b)
J. Khoury, A. Weltman, Phys. Rev. Lett. **93**, 171104 (2004a)
J. Khoury, A. Weltman, Phys. Rev. D **69**, 044026 (2004b)
J.A. King et al., Mon. Not. R. Astron. Soc. **422**, 3370 (2012)
A. Konopliv et al., Icarus **211**, 401 (2011)
A. Kostelecky, M. Mewes, Phys. Rev. D **66**, 056005 (2002)
A. Kostelecky, N. Russel, Rev. Mod. Phys. **83**, 11 (2011)
S. Lambert, C. Le Poncin-Lafitte, Astron. Astrophys. **529**, A70 (2011)
S. Lamoreaux et al., Phys. Rev. Lett. **57**, 3125 (1986)
C. Le Poncin-Lafitte et al., arXiv:1604.01663 (2016)
N. Leefer et al., Phys. Rev. Lett. **111**, 060801 (2013)
D. Lovelock, J. Math. Phys. **12**, 498 (1971)
D. Lovelock, J. Math. Phys. **13**, 874 (1972)
R. Mansouri, R. Sexl, Gen. Relativ. Gravit. **8**, 497 (1977)
C.J.A.P. Martins, A.M.M. Pinho, Phys. Rev. D **91**, 103501 (2015)
D. Mattingly, Living Rev. Relativ. **8**, 5 (2005)
P.J. McMillan, Mon. Not. R. Astron. Soc. **414**, 2446 (2011)
S. Merlet et al., Metrologia **47**, L9 (2010)
F. Mignard, S. Klioner, Proc. IAU Symp. **261**, 306 (2010)
A. Milani et al., Phys. Rev. D **66**, 082001 (2002)
M. Milgrom, Astrophys. J. **270**, 365 (1983a)
M. Milgrom, Astrophys. J. **270**, 371 (1983b)
M. Milgrom, Astrophys. J. **270**, 384 (1983c)

M. Milgrom, Mon. Not. R. Astron. Soc. **399**, 474 (2009)
O. Minazzoli, A. Hees, Phys. Rev. D **90**, 023017 (2014)
N. Mohapi et al., J. Cosmol. Astropart. Phys. **3**, 32 (2016)
O. Montenbruck et al., GPS Solut. **16**(3), 303–313 (2012)
O. Montenbruck et al., GPS World **24**(7), 44–49 (2013)
O. Montenbruck et al., J. Geod. **89**(3), 283–297 (2014)
S. Mouret, Phys. Rev. D **84**, 122001 (2011)
H. Müller et al., Phys. Rev. Lett. **100**, 031101 (2008)
M.T. Murphy et al., Mon. Not. R. Astron. Soc. **345**, 609 (2003)
R.M. Neal, Ann. Stat. **31**(3), 705–767 (2003)
N. Nunes, J. Lidsey, Phys. Rev. D **69**, 123511 (2004)
S. Origlia et al., arXiv:1603.06062 (2016)
J. Overduin, P. Wesson, Phys. Rep. **283**, 303 (1997)
S. Peil et al., Phys. Rev. A **87**, 010102(R) (2013)
P. Perez, Y. Sacquin, Class. Quantum Gravity **29**, 184008 (2012)
A. Peters et al., Nature **400**, 849 (1999)
A. Peters et al., Metrologia **38**, 25 (2001)
E. Pitjeva, N. Pitjev, Mon. Not. R. Astron. Soc. **432**, 3431 (2013)
Planck Collaboration, Astron. Astrophys. **571**, A16 (2013)
R. Pound, G. Rebka, Phys. Rev. Lett. **3**, 439 (1959)
L. Prange et al., in *IAG Potsdam 2013 Proceedings. International Association of Geodesy Symposia*, 2014, ed. by P. Willis (Springer, Berlin, 2013)
B. Ratra, P. Peebles, Phys. Rev. D **37**, 3406 (1988)
S. Reinhardt et al., Nat. Phys. **3**, 861 (2007)
H. Robertson, Rev. Mod. Phys. **21**, 378 (1949)
P. Rochat et al., in *Proceedings of 2012 European Navigation Conference*, Gdansk, Poland, 25–27 Apr. 2012
T. Rosenband et al., Science **319**, 1808 (2008)
G. Saathoff et al., Phys. Rev. Lett. **91**, 190403 (2003)
S. Schlamminger et al., Phys. Rev. Lett. **100**, 041101 (2008)
L. Shao, Phys. Rev. Lett. **112**, 111103 (2014)
R. Srianand et al., Phys. Rev. Lett. **92**, 121302 (2004)
Y. Stadnik, V. Flambaum, Phys. Rev. Lett. **114**, 161301 (2014)
Y. Stadnik, V. Flambaum, Phys. Rev. Lett. **115**, 201301 (2015)
P. Stanwix et al., Phys. Rev. D **74**, 081101(R) (2006)
STE-QUEST, Assessment study report. ESA report ESA/SRE 6 (2013). Available on http://sci.esa.int/ste-quest/53445-ste-quest-yellow-book/
C. Stubbs, Phys. Rev. Lett. **70**, 119 (1993)
R. Swaters et al., Astrophys. J. **718**, 380 (2010)
C. Talmadge et al., Phys. Rev. Lett. **61**, 1159 (1988)
M. Tarallo et al., Phys. Rev. Lett. **113**, 023005 (2014)
J. Tasson, Rep. Prog. Phys. **77**, 6 (2014)
K. Thorne, D. Lee, A. Lightman, Phys. Rev. D **7**, 3563 (1971)
S. Turyshev, Phys. Usp. **52**, 1 (2009)
J.-P. Uzan, Living Rev. Relativ. **14**, 2 (2011)
K. Van Tilburg et al., Phys. Rev. Lett. **115**, 011802 (2015)
A. Verma et al., Astron. Astrophys. **561**, A115 (2014)
R.F.C. Vessot, M.W. Levine, Gen. Relativ. Gravit. **10**, 181–204 (1979)
R.F.C. Vessot et al., Phys. Rev. Lett. **45**, 2081 (1980)
R.F.C. Vessot et al., Adv. Space Res. **9**, 21–28 (1989)
J. Wang, L. Hui, J. Khoury, Phys. Rev. Lett. **109**, 241301 (2012)
J. Webb et al., Phys. Rev. Lett. **87**, 091301 (2003)
J. Webb et al., Phys. Rev. Lett. **107**, 191101 (2011)
C. Will, *Theory and Experiment in Gravitational Physics* (Cambridge University Press, Cambridge, 1993)
C. Will, Living Rev. Relativ. **17**, 4 (2014)
J. Williams et al., Int. J. Mod. Phys. D **18**, 1129 (2009)
J. Williams et al., Class. Quantum Gravity **29**, 184004 (2012)
P. Wolf, G. Petit, Phys. Rev. A **56**, 4405 (1997)
P. Wolf et al., Phys. Rev. Lett. **90**, 060402 (2003)
P. Wolf et al., Phys. Rev. Lett. **96**, 060801 (2006a)
P. Wolf et al., Phys. Rev. Lett. **96**, 060801 (2006b)
L. Zhou et al., Phys. Rev. Lett. **115**, 1 (2015)
I. Zlatev et al., Phys. Rev. Lett. **82**, 896 (1999)
S. Zucker et al., Astrophys. J. **639**, L21 (2006)

Space Sci Rev (2018) 214:3
DOI 10.1007/s11214-017-0432-y

Search for Varying Constants of Nature from Astronomical Observation of Molecules

Wim Ubachs[1]

Received: 20 April 2017 / Accepted: 10 October 2017 / Published online: 30 November 2017

Abstract The status of searches for possible variation in the constants of nature from astronomical observation of molecules is reviewed, focusing on the dimensionless constant representing the proton-electron mass ratio $\mu = m_p/m_e$. The optical detection of H_2 and CO molecules with large ground-based telescopes (as the ESO-VLT and the Keck telescopes), as well as the detection of H_2 with the Cosmic Origins Spectrograph aboard the Hubble Space Telescope is discussed in the context of varying constants, and in connection to different theoretical scenarios. Radio astronomy provides an alternative search strategy bearing the advantage that molecules as NH_3 (ammonia) and CH_3OH (methanol) can be used, which are much more sensitive to a varying μ than diatomic molecules. Current constraints are $|\Delta\mu/\mu| < 5 \times 10^{-6}$ for redshift $z = 2.0$–4.2, corresponding to look-back times of 10–12.5 Gyrs, and $|\Delta\mu/\mu| < 1.5 \times 10^{-7}$ for $z = 0.88$, corresponding to half the age of the Universe (both at 3σ statistical significance). Existing bottlenecks and prospects for future improvement with novel instrumentation are discussed.

Keywords Varying constants · Extragalactic astronomy · Molecular spectroscopy

1 Introduction

Extensions of the Standard Model (SM) of physics can occur in various ways. One possibility of such an extension is the temporal or spatial variation of fundamental constants of nature, such as the coupling strengths of the known four forces, the strong and weak forces, electromagnetism and gravity. Their coupling strengths are inserted in the SM as

High Performance Clocks with Special Emphasis on Geodesy and Geophysics and Applications to Other Bodies of the Solar System
Edited by Rafael Rodrigo, Véronique Dehant, Leonid Gurvits, Michael Kramer, Ryan Park and Peter Wolf

✉ W. Ubachs
w.m.g.ubachs@vu.nl

[1] Department of Physics and Astronomy, Vrije Universiteit Amsterdam, De Boelelaan 1081, 1081 HV Amsterdam, The Netherlands

free parameters, as are the relative masses of the elementary particles. However, in order to describe physics in a consistent framework of quantum field theories it does not suffice to allow the fundamental constants to vary at will. For consistency, and to retain e.g. conservation of energy, the inclusion of additional fields is required, often referred to as 'dilaton' fields that may in the simplest form be of scalar nature (Barrow et al. 2002). Such fields may be regarded as additional forces or 'quintessence', and may be associated with other unknown forms of (dark) matter or energy. If such connection exists, then the values of some constants of nature might follow the increasing amount of dark energy in the Universe, in which case the variation of constants would manifest itself on a cosmological time scale. Alternatively, the fundamental constants might be connected to local effects such as matter density (Khoury and Weltman 2004) or gravitational fields (Flambaum and Shuryak 2008). Theories involving variation of fundamental constants exist in various flavors and are e.g. described in the review by Uzan (2011).

The quantum mechanical energy level structure of atoms and molecules is determined by the value of the fine structure constant α and the proton-to-electron mass ratio $\mu = m_p/m_e$ (Kozlov and Levshakov 2013; Jansen et al. 2014). Searches for variation of both dimensionless constants α and μ focus primarily on spectroscopic methods. Comparison between spectra obtained from astronomical observations at high redshift with the same spectra measured at high precision in the present epoch with laboratory-based methods may reveal minute shifts in the values of those constants. Spectroscopy, i.e. the measurement of wavelengths or frequencies of absorption lines, is a solid method in view of the accuracy at which such quantities can be measured. In the analysis procedures of comparing spectra the small differential shifts of absorption lines imposed by varying constants must be disentangled from the cosmological and gravitational redshifts involved in astronomical observations. However, General Relativity (GR) predicts that these latter redshifts are dispersionless, hence disentangling is straightforward. As variation of constants implies a breakdown of Einstein's equivalence principle, a corner stone of existing theories of physics, any resulting differential or dispersion effect may be considered as a token for new physics.

The search for a varying fine structure constant α from comparison of spectroscopic lines was pursued already in the 1950s (Savedoff 1956) based on the alkali-doublet method. Here it is used that the separation of two spectral lines probing a spin-orbit doublet from a common ground state scales with α^2. Bahcall and Schmidt (1967) therewith constrained a relative variation of α at the 10^{-3} level from observations of a pair O III emission lines at redshift $z = 0.2$. The accuracy was much improved by the Many-Multiplet-(MM) method, where absolute transition frequencies of various elements can be used simultaneously for probing a variation of α. The MM-method requires detailed ab initio calculations of atomic structure to derive sensitivity coefficients (Dzuba et al. 1999). Such sensitivity coefficients reflect by how much spectral lines will shift as a result of a relative change in the fine structure constant $\Delta\alpha/\alpha$, and they appear to be largest for heavy elements where electrons undergo relativistic motion. Based on the MM-method and observations with large 8–10 m class telescopes certain indications of a temporally (Webb et al. 2001) and spatially (Webb et al. 2011) varying α were reported. It is discussed in the scientific literature whether these effects might be due, at least in part, to calibration issues with the spectrographs connected to the large optical telescopes used to probe varying α at high redshift (Rahmani et al. 2013; Whitmore and Murphy 2015; Dumont and Webb 2017).

Focus of this paper is on another dimensionless constant, the proton-to-electron mass ratio μ, and the constraints that can be derived on its relative variation $\Delta\mu/\mu$. In theoretical models based on Grand Unification it was shown that the rate of change in μ is much faster than a putative rate of change in α (Calmet and Fritzsch 2002; Langacker et al. 2002), thus

Table 1 List of the ten high redshift H_2 absorption systems analyzed so far for μ-variation. The identification of the quasar source, the redshift z of the absorbing system, and the J2000-coordinates of the quasar source are listed. The column densities $N(H_2)$ are given on a $\log_{10}$ scale in cm^{-2}. R_{mag} represents the magnitude in the visible range

Quasar	z_{abs}	RA(J2000)	Decl.(J2000)	$N(H_2)$	R_{mag}	Refs.
HE0027-1836	2.42	00:30:23.62	−18:19:56.0	17.3	17.37	Rahmani et al. (2013)
Q0347-383	3.02	03:49:43.64	−38:10:30.6	14.5	17.48	Ivanchik et al. (2005), Reinhold et al. (2006), King et al. (2008), Wendt and Molaro (2012), Thompson et al. (2009)
Q0405-443	2.59	04:07:18.08	−44:10:13.9	18.2	17.34	Ivanchik et al. (2005), Reinhold et al. (2006), King et al. (2008), Thompson et al. (2009)
Q0528-250	2.81	05:30:07.95	−25:03:29.7	18.2	17.37	King et al. (2008, 2011)
B0642-5038	2.66	06:43:26.99	−50:41:12.7	18.4	18.06	Bagdonaite et al. (2014b), Albornoz Vásquez et al. (2014)
Q1232+082	2.34	12:34:37.58	+07:58:43.6	19.7	18.40	Varshalovich et al. (2001), Ivanchik et al. (2010), Daprá et al. (2017b)
J1237+064	2.69	12:37:14.60	+06:47:59.5	19.2	18.21	Daprà et al. (2015)
J1443+2724	4.22	14:43:31.18	+27:24:36.4	18.3	18.81	Bagdonaite et al. (2015)
J2123-0050	2.06	21:23:29.46	−00:50:52.9	17.6	15.83	Malec et al. (2010), van Weerdenburg et al. (2011)
Q2348-011	2.42	23:50:57.87	−00:52:09.9	18.4	18.31	Bagdonaite et al. (2012)

making μ a sensitive test ground for probing varying constants. Typically the constant μ is associated with the quantized motion of nuclei in molecules, through their intramolecular rotation, vibration and tunneling modes. Again for any model comparing astronomical spectra with zero-redshift laboratory data, sensitivity coefficients $K_i = (d \ln \lambda)/(d \ln \mu)$ must be evaluated for each allowed quantum transition in a molecule (Jansen et al. 2014). In the following the results of ground-based observation of molecular hydrogen and carbon monoxide at high-redshift will be presented, yielding constraints on $\Delta\mu/\mu$. It will further be discussed how vacuum-ultraviolet (VUV) absorption by H_2 in galactic white dwarf objects, detected by a satellite-based spectrograph, will be used for constraining a possible dependence of μ on a gravitational field. The microwave detection of molecular absorption used for probing a varying μ is described in a subsequent section. Future developments in optical and radio astronomy, as well as bottlenecks, challenges and opportunities for more tight constraints on variation of the proton-electron mass ratio will be discussed in a concluding section.

2 Optical Detection of the H_2 Molecule

Hydrogen is by far the most abundant species in Galaxies and is found in both atomic and molecular forms in the interstellar medium. The investigation of H_2 absorption in galaxies at high redshift in the line-of-sight towards quasars favorably draws from this omnipresence of H_2. Nevertheless, although large numbers of quasars have been detected, many of which with a Damped-Lyman-α (DLA) absorption feature marking a column density for atomic hydrogen in excess of 2×10^{20} cm^{-2}, only a limited number of systems with pronounced H_2 absorption have been identified. Currently some 23 H_2 absorbers at redshifts $z > 2$ have been found by various authors and a compilation can be found in Ubachs et al. (2016). An additional set of 23 systems with tentative absorptions have been detected (Balashev et al. 2014). In those systems typically a column density for H_2 is detected at the 1% level with respect to that of H, although much higher molecular fractions have also been detected recently, see e.g. Noterdaeme et al. (2017).

The strongest electric dipole-allowed transitions in the H_2 molecule, those of the Lyman and Werner band systems, occur at rest-frame wavelengths of 912–1150 Å, so that a significant fraction of the absorption lines shift into the atmospheric transmission window ($\lambda > 3050$ Å) for absorption redshifts of $z > 2$. Since the molecular absorption bands lay shortward in wavelength of the Lyman-α line (1215 Å), H_2 line absorptions are typically found in the 'Lyman-α forest'. This consists of a spectral region crowded with many Lyman-α absorption lines, produced by the intergalactic gas over the full range of redshifts along the line of sight, from the observer to the background quasar (Rauch 1998). This Lyman-α forest produces in fact a random spectrum overlaying the spectral region where H_2 absorption lines are located, thus forming an obstacle to be overcome in the data analysis (Malec et al. 2010).

Currently some ten H_2 absorptions systems, listed in Table 1, have been analyzed for μ-variation. These are good quality systems in view of brightness of the background quasar, typically in the range of $R_{mag} = 17.3$–19. Such magnitude allows for recording of an H_2 absorption spectrum of typical SNR ≥ 20 and resolution of $R = 50\,000$ in an amount of 10 hrs at an 8–10 m class telescope, such as the Keck telescope (with the HIRES spectrograph) and the Very Large Telescope (VLT, with the UVES spectrograph). The quasar J2123-005 is a special case displaying extraordinary high brightness ($R_{mag} = 15.83$), which was exploited in observations by both telescopes (Malec et al. 2010; van Weerdenburg et al. 2011). Further, good quality H_2 absorption spectra can be obtained for column densities of $\log_{10} N(H_2)$ [cm^{-2}] $= 17$–18.5. System Q0347-383 with $\log_{10} N(H_2)$ [cm^{-2}] $= 14.5$ produces nevertheless a good quality spectrum with non-saturated lines (Ivanchik et al. 2005; Wendt and Molaro 2012), while system J1237+064 exhibits strongly saturated absorption lines for the highest populated quantum states of H_2 (Noterdaeme et al. 2010).

Quasar absorption spectra are treated in a 'comprehensive fitting' analysis (Malec et al. 2010), where all H_2 lines are described by a set of molecular parameters, λ_i^0 for the rest-frame wavelength, f_i for the line oscillator strength, and Γ_i for the natural broadening coefficient, and the sensitivity coefficients K_i and invoked in the equation governing the comparison between spectra:

$$\lambda_i^z = \lambda_i^0 (1 + z_{\text{abs}}) \left(1 + K_i \frac{\Delta\mu}{\mu} \right). \tag{1}$$

The rest frame wavelengths λ_i^0 of the H_2 absorption lines have been determined to high accuracy in dedicated laboratory experiments using tunable vacuum ultraviolet lasers (Ubachs and Reinhold 2004; Philip et al. 2004), as well as by a combination of spectroscopic studies aimed at determining level energies of excited states in H_2 from which the vacuum ultraviolet absorption lines can be calculated at high accuracy (Bailly et al. 2010). Also lines of the HD isotopologue are used in the astronomical analyses for which accurate laboratory data are available (Ivanov et al. 2008). The status of these laboratory experiments is that the transition wavelengths of the H_2 absorptions may be considered exact for the purpose of comparison to quasar data recorded with the HIRES spectrometer at Keck and the UVES

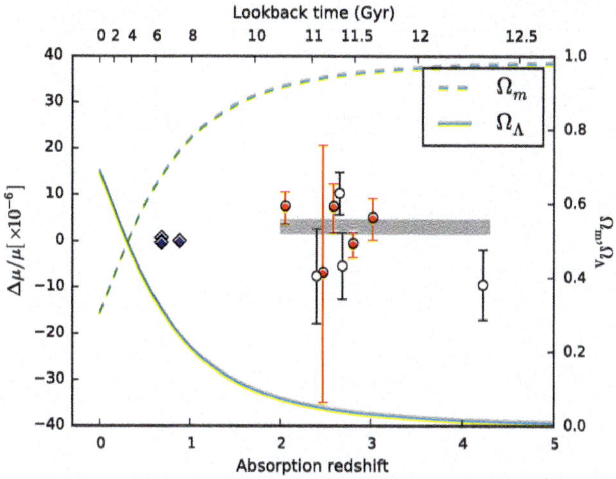

Fig. 1 Values for the relative variation of the proton-to-electron mass ratio $\Delta\mu/\mu$ plotted as a function of redshift and look-back time for nine H_2 absorption systems analyzed in detail (Q1232+082 not included). *Open circles*: data corrected for long-range wavelength distortions; *closed (red) circles*: uncorrected data. Results are plotted alongside with evolution of cosmological parameters for Ω_m (*dashed line*) and Ω_Λ (*full line*), with respect to the right-hand vertical axis. The *gray bar* represents the average for $\Delta\mu/\mu$ with $\pm 1\sigma$ uncertainty limits from the H_2 systems analyzed. For comparison results from radio astronomical observations at lower redshift ($z < 1$) are plotted as well (*diamonds—blue*). Figure reproduced from Ubachs et al. (2016)

spectrometer at VLT. The values f_i and Γ_i for the absorption lines are obtained from ab initio calculations of the H_2 molecule (Abgrall et al. 2000). Values for the sensitivity coefficients K_i were derived by a number of groups using independent techniques (Varshalovich and Levshakov 1993; Meshkov et al. 2006; Ubachs et al. 2007).

In the analysis of astronomical spectra each spectral line is folded by a Doppler parameter b, shifted by the redshift parameter $1 + z$ and its intensity multiplied by the column density N_J for each populated quantum state of the H_2 molecule. Typically only states with $v = 0$ and $J = 0$–5 will be populated to the extent that absorption is detected. Hence, from the broadening parameters Γ_i and b, further folded by the instrument function of the spectrograph, a Voigt profile shape for the lines is obtained. While applying the fitting routine to the entire spectrum multiple velocity components, introduced as multiple redshift parameters z_i, will be allowed for optimizing the entire spectrum. In this analysis the occurrence of Lyman-α forest lines as well as metal line absorption will be included. In the final stage of this comprehensive fitting method, after a well-behaved absorption model is created, $\Delta\mu/\mu$ is introduced as a final free parameter and fitted.

This procedure has been applied to several of the ten absorption systems listed in Table 1, while for some other systems a 'line-by-line' fitting model was used (Thompson et al. 2009; Wendt and Molaro 2012). The latter method exhibits some drawbacks when dealing with overlapped lines in the fit (Malec et al. 2010; King et al. 2011). Results for the values $\Delta\mu/\mu$ obtained from the various systems at redshifts in the range $z = 2.0$–4.2 analyzed have been collected in Fig. 1. A weighted average yields $\Delta\mu/\mu = (3.1 \pm 1.6) \times 10^{-6}$. This result, indicating a small 1.9σ effect of a varying μ is, in view of its small significance, treated as a null-result. This null-result may also be phrased as $|\Delta\mu/\mu| < 5 \times 10^{-6}$ at a 3σ significance level (Ubachs et al. 2016).

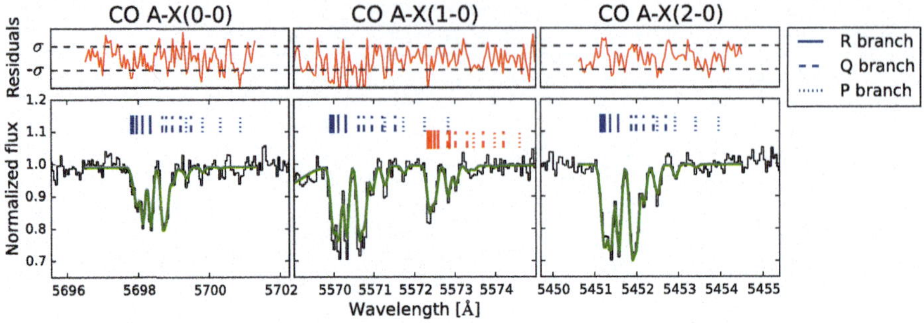

Fig. 2 Spectra of the A-X $(v, 0)$ bands, for $v = 0$–2, observed in the line of sight toward J1237+064 at redshift $z = 2.69$. In the *upper panels* the residuals from a fit are displayed

3 Optical Detection of the CO Molecule

Carbon monoxide is the second most abundant gaseous molecule in the Universe and its A-X electronic band system, occurring at rest-frame wavelengths of $\lambda < 1544$ Å, is observed in a number of quasar absorption systems (Noterdaeme et al. 2008, 2010, 2011, 2017). Since the wavelength interval of the A-X bands is longward of the Lyman-α line these CO spectra are not overlaid by Lyman-α forest lines.

In the quasar absorption system J1237+064 CO is detected at $z = 2.69$ (with ESO-VLT) with a column density of $\log_{10} N(\mathrm{CO})$ [cm^{-2}] $= 14.2$ and absorption bands A-X$(v, 0)$ for $v = 0$–8 were observed in the range $\lambda = 5696$–4880 Å. In addition also the B-X(0, 0), C-X(0, 0) and E-X(0, 0) bands were observed in the bluer part of the spectrum. Molecular parameters were measured in laser-based and synchrotron-based spectroscopic absorption studies (Salumbides et al. 2012) and invoked in the 'comprehensive fitting' method based on Eq. (1). In Fig. 2 spectra of the three lowest bands of CO A-X, as observed in J1237+064, are displayed with a model fit. The analysis of the CO spectrum yields a value of $\Delta\mu/\mu = (2.3 \pm 1.7) \times 10^{-5}$ (Daprà et al. 2016). This shows that the CO method is somewhat less constraining than the H_2 method, which is ascribed to the fact that the details of the CO absorption features are spread over less pixels in the spectrum than those of H_2. The range of K-coefficients is similar for both molecules. Recently also μ-variation was analyzed from a second absorber, J0000+0048 at $z = 2.52$, resulting in $\Delta\mu/\mu = (1.8 \pm 2.3) \times 10^{-5}$ (Daprá et al. 2017a).

4 Detection of H_2 in the Photosphere of White Dwarfs

The spectrum of molecular hydrogen can also be used to search for a possible dependence of the fundamental constant μ on a gravitational field. Theoretically such phenomenon might be explained in 'chameleon' scenarios of varying constants (Khoury and Weltman 2004; Flambaum and Shuryak 2008). Some white dwarfs (WD) exhibit a thin photosphere layer in which molecular hydrogen molecules give rise to an absorption spectrum (Xu et al. 2013). The conditions applying to the H_2 molecules in such circumstances are extreme. Since white dwarfs are collapsed stars of typically a solar mass shrunk to a radius R of 1% of the Sun, the gravitational potential $\phi = GM/Rc^2$ is about 10 000 times that on the Earth's surface. Further, the temperature of the photosphere is typically in excess of 10 000 K. Spectra of the photosphere of two white dwarfs, GD-29-38 (or WD2326+049) and GD-133

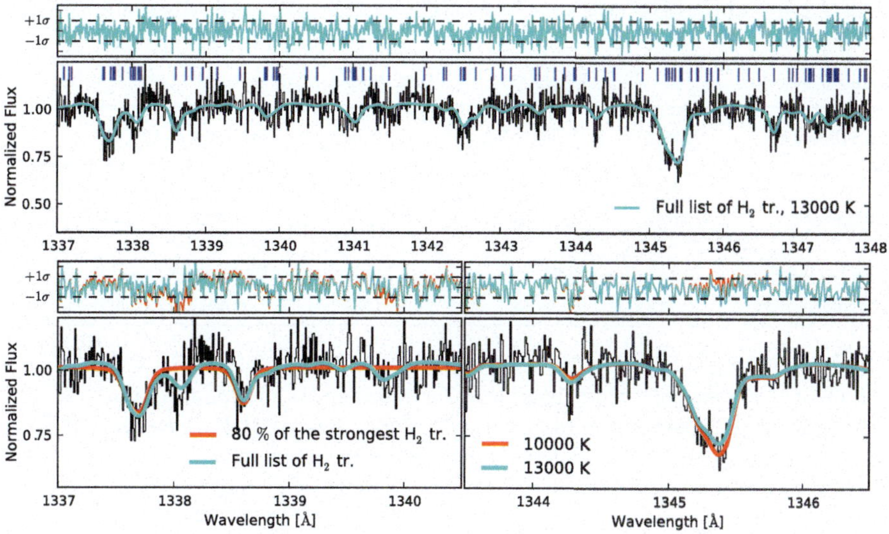

Fig. 3 Part of the absorption spectrum of the photosphere of the white dwarf system GD29-38, overlaid with a model spectrum for temperatures of $T = 10\,000$ K and using the 80% strongest lines (*red line*), and a model for $T = 13\,000$ K and the full list of H_2 absorption lines (*blue-green line*). The total spectrum used in the analysis extends from 1310–1412 Å

(WD1116+026) were observed with the Cosmic Origins Spectrograph (COS) aboard the Hubble Space Telescope (HST).

Studies aiming to investigate gravitational effects in dim objects as WD stars can preferably be performed on systems observed inside the Galaxy. However, in such case there will be no redshift to shift the absorption profiles into the atmospheric transmission window. Galactic WD spectra should then be recorded in the vacuum ultraviolet range, for which COS-HST is ideally suited. An example of a Galactic WD H_2 absorption spectrum is displayed in Fig. 3 for the case of GD-29-38. In view of the high local temperatures higher vibrational states of H_2 are populated in the WD photospheres, in contrast to the cold environments probed in quasar absorption, where only H_2 ($v = 0$) is detected. In view of the Franck-Condon factors of H_2 the Lyman bands B-X(1, 3) and (1, 4) are the most strongly represented in the observed spectrum which covers the range 1310–1412 Å (Bagdonaite et al. 2014a). At temperatures of $T > 10\,000$ K rotational quantum states of up to $J = 25$ are effectively populated and a simulation of the spectra involves starting from the partition function including all states populated (Salumbides et al. 2015):

$$P_{v,J}(T) = \frac{g_n(2J+1)e^{\frac{-E_{v,J}}{kT}}}{\sum_{v=0}^{v_{\max}} \sum_{J=0}^{J_{\max}(v)} g_n(2J+1)e^{\frac{-E_{v,J}}{kT}}} \tag{2}$$

Based on this partition function and assuming thermodynamic equilibrium for the H_2 populated quantum states a fit can be made to the observed spectra, with temperature T and $\Delta\mu/\mu$ as free parameters, again using Eq. (1) for the absorption model. This method results in constraints on dependences of μ on a gravitational field quantified as $\Delta\mu/\mu = (-5.8 \pm 3.9) \times 10^{-5}$ for GD-29-38 (at $\phi = 1.9 \times 10^{-4}$) and $\Delta\mu/\mu = (-2.7 \pm 4.9) \times 10^{-5}$ for GD-133 (at $\phi = 1.2 \times 10^{-4}$) (Bagdonaite et al. 2014a). Thus far the objects GD-29-38

Table 2 List of extra-galactic radio sources with molecules detected in absorption searching for varying constants

Radio Quasar	z_{abs}	RA(J2000)	Decl.(J2000)	Refs.
PKS1830-211	0.882	18:33:39.9	−21:03:40	Henkel et al. (2009), Bagdonaite et al. (2013a, 2013b), Kanekar et al. (2015)
B0218+357	0.684	02:21:05.5	+35:56:14	Murphy et al. (2008), Kanekar (2011)
PKS1413+135	0.247	14:15:58.8	+13:20:24	Darling (2004)
PMN0134-0931	0.765	01:34:35.7	−09:31:03	Kanekar et al. (2005)

and GD-133 are the only white dwarf stars for which an H_2 absorption spectrum of sufficient SNR was obtained. In a third system, GD-31 (Xu et al. 2013), H_2 was observed but the spectrum is of too low quality to derive a competitive constraint on μ.

5 Radio Astronomical Observation of Polyatomic Molecules

Radio astronomy is the preferred technique for identifying molecules in the interstellar medium and currently well over a 100 molecular species have been detected through observations in our Galaxy. Only a limited sample of extragalactic radio sources at intermediate redshift ($0.2 < z < 1$) with molecular absorption in its sightline have been identified, the four listed in Table 2 which have been used to search for varying constants on a cosmological time scale, and a fifth system B1504+377 in which OH has been detected (Kanekar and Chengalur 2002). The radio-loud quasars B0218+357 and PKS 1830-211 are strongly lensed radio galaxies with the radio source far in the background, exhibit a large extinction, $A_v \sim 10$–100, and thus a very high molecular fraction.

Most of the molecules detected in the radio range are observed through pure rotational lines in the spectrum, for which the sensitivity coefficients are $K_\mu = -1$ (Jansen et al. 2014). This implies that all such transitions exhibit the same sensitivity to μ-variation. Hence μ-variation does not produce a differential effect, to be distinguished from a cosmological or a Doppler shift, not even if transitions of different molecules are included in a combined analysis. If however molecules are considered that exhibit other internal modes of energy a favorable situation arises. The ammonia molecule exhibits an inversion splitting associated with a tunneling inversion mode (the N-atom tunneling through the plane of the three H-atoms) which is very sensitive to a mass variation, in view of quantum tunneling being exponentially dependent on mass. This gives rise to a sensitivity coefficient of $K_\mu = -4.2$ for the typical 21 GHz lines of NH_3 (Flambaum and Kozlov 2007). However, again the various rotational-tunneling lines all exhibit the same K-value, so that purely rotational lines of other molecules (at $K_\mu = -1$) must be included to derive constraints on $\Delta\mu/\mu$. Typically HC_3N lines can be used for that purpose (Henkel et al. 2009), but the analysis then is based on the assumption that NH_3 and HC_3N molecules reside in the same location in the extragalactic absorbing clouds. Ammonia has thus far been observed in two intermediate redshift objects, PKS1830-211 (Henkel et al. 2009) and B0218+357 (Murphy et al. 2008).

The methanol molecule is an exemplary case where these shortcomings are circumvented. The motion of hindered internal rotation of the OH-group over the CH_3 group gives rise to a tunneling motion for which some rotational-torsional transitions exhibit very large K_μ coefficients, with values of $K_\mu = -7.4$ and $K_\mu = -33$ (Jansen et al. 2011; Levshakov

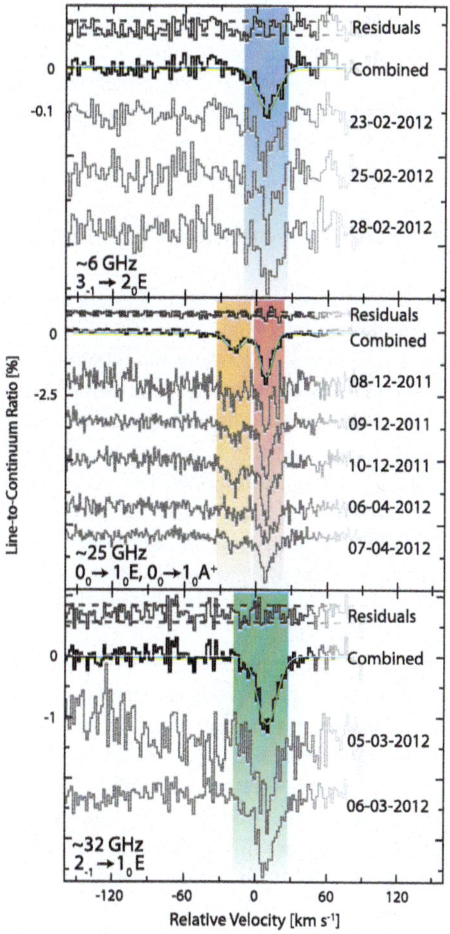

Fig. 4 The four methanol absorption lines observed with the Effelsberg radio telescope at $z = 0.88582$ towards the radio-loud quasar PKS1830-211. Figure adapted from Bagdonaite et al. (2013b)

et al. 2011). When these transitions are combined with pure rotational transitions, of the same molecular species, then a very sensitive analysis of μ-variation can be performed. This was done in an observation of four radio lines in the methanol spectrum, ranging from 6.5 GHz to 32 GHz, with the Effelsberg 100 m dish radio telescope (see Fig. 4; Bagdonaite et al. 2013b). After that initial study more extended observations of methanol in the same object have been observed, using also the IRAM 30 m telescope and the ALMA telescope array (Bagdonaite et al. 2013a), as well as the Extended Very Large Array (Kanekar et al. 2015). These studies have produced the tightest constraint of $|\Delta\mu/\mu| < 10^{-7}$, however only for a single object and at a single redshift, equivalent to 7.5 Gyrs look-back time or half the age of the Universe. The stringent constraints on $\Delta\mu/\mu$ derived from NH_3 and CH_3OH at redshifts $z = 0.68$ and $z = 0.88$ are included in Fig. 1.

In two other radio systems PMN J0134-0931 with a lensed galaxy at $z = 0.765$, and PKS1413+135 absorption of OH radicals was analyzed that provided combined constraints on μ and α. The observation of OH lines at 18 cm was combined with that of the HI 21 cm line, from which contributions to varying μ and α may be disentangled (Darling 2004; Kanekar et al. 2005).

Most investigations on varying constants focus on absorption, because cold environments may be probed without too much turbulent motion and line broadening. It was shown however, that emission of CO combined with emission of atomic C I, towards the lensed galaxy HLSJ091828.6+514223 at $z = 5.2$ could be employed to derive a constraint on μ-variation at such high redshifts (Levshakov et al. 2012).

6 Future Opportunities and Instrumentation

The accuracy of the H_2 and CO optical absorption methods as discussed in Sects. 2 and 3 is limited by the signal-to-noise ratio (SNR) obtained in the spectra recorded at the 8–10 m class telescopes. Extending the integration time will improve the spectra and therewith the statistical constraint on $\Delta\mu/\mu$, but the SNR scales only with the root of measurement time. Since for the measurement of the H_2 absorption spectrum of Q2348-011 already an averaging time of 19 hrs on target was applied (Bagdonaite et al. 2012), it seems unrealistic to further extend such long observation periods on 8–10 m class telescopes. However, the construction of a next generation optical telescopes with 30–40 m dishes, such as the European Extremely Large Telescope, the Thirty Meter Telescope and the Giant Magellan Telescope, will deliver the prospect of increased photon collection and drastically improved signal-to-noise ratio. This will straightforwardly lead to tighter bounds on $\Delta\mu/\mu$ with the H_2 and CO optical methods.

The spectrographs used for μ-variation studies (UVES at VLT, and HIRES at Keck) are designed such that a resolving power of $R \sim 50\,000$ is obtained for an entrance slit corresponding to 0.8 arcseconds; this value matches 'regular-to-good' seeing conditions so that all light can be captured within the slit at seeing conditions of $0.8''$. For the molecular hydrogen absorbing systems investigated so far it was found that the Doppler broadening of the individual H_2 features in all of the cases yielded superthermal line broadenings in excess of the 50 000 resolving power limit, most likely due to turbulent motion inside the absorbing galaxy or unresolved velocity substructures. In the specific case of J2123-005 spectra were measured at $R = 55\,000$ with UVES-VLT (van Weerdenburg et al. 2011) and $R = 110\,000$ with HIRES-Keck (Malec et al. 2010), but the linewidths in the observed spectra were the same, hence fully governed by superthermal Doppler broadening. In such cases, for all systems investigated (see Table 1), the effective Doppler width greatly exceeds the kinematic width that would correspond to the Boltzmann population temperatures, which are for H_2 typically 50–150 K (Petitjean et al. 2002; Srianand et al. 2005). Although it might be possible that an H_2 absorbing system will be found, exhibiting no turbulent motions and a strongly decreased Doppler broadening, as of now it appears that improved resolution beyond 50 000 would probably not benefit the H_2 method. The planned installation of the ESPRESSO spectrograph at ESO-VLT with its high-resolution mode of $R = 120\,000$ and its very-high-resolution mode of $R = 220\,000$ would provide an opportunity to further test this hypothesis.

Wavelength calibration is a crucial issue in a spectroscopic search for varying constants, because the methods rely on exploring the smallest differential wavelength shifts between molecular absorption lines. The spectra are generally calibrated against a ThAr reference spectrum measured immediately after the quasar exposures without resetting its geometry. It was found that "intra-order" distortions occur in the echelle-grating based measurements (Griest et al. 2010; Whitmore et al. 2010). In addition, it was found that long-range wavelength distortions also affect the calibration of the quasar absorption spectra when calibrated against the ThAr lines. This is probably due to the fact that the light beams as originating from far-distant objects and the light beams from the ThAr cell on the telescope platform travel along different paths (Rahmani et al. 2013; Whitmore and Murphy 2015). The resulting distortions of the wavelength scale may mimic a $\Delta\mu/\mu$ effect. For analyzing these miscalibration phenomena exposures of either solar twin stars or of asteroids are used, which may be compared with an accurate high-resolution Fourier-transform spectra measured of the Sun (Molaro et al. 2008). Such recalibration methods may be applied to assess a value for the systematic uncertainty in the determination of $\Delta\mu/\mu$, or even counter-distort the wavelength scale to obtain a fiducial value of a constraint on $\Delta\mu/\mu$, as was pursued in some studies (Rahmani et al. 2013; Bagdonaite et al. 2014b; Daprà et al. 2015).

Some of the problems of wavelength calibration of astronomical spectra can be resolved by fiber-feeding the light into the spectrograph. Already for the ESPRESSO spectrograph the light from the VLT large-dish will be collected into a multi-mode fiber before entering the spectrograph, with the goal to discriminate against angular differences between light paths. A problem with fiber-feeding is the additional loss of photons, in particular in the UV part of the spectrum. In fact the fiber-fed spectrographs will not be specified for the range 3000–3800 Å, a range that contains the central information for the H_2 method in case of quasars at redshifts $z = 2$–3. Hence fiber feeding the spectrograph will be detrimental in application to most of the H_2 absorbers that fall in that redshift range (see Table 1).

Another improvement, currently established on some telescope-based spectrographs, is the use of frequency comb lasers at the observatories (Wilken et al. 2012). Frequency comb lasers deliver a very dense spectrum of frequency markers at a separation exactly governed by an atomic clock, thus constituting an ideal reference spectrum. However, extension of the comb markers to the UV part of the spectrum at 3000–3800 Å is not obvious. Again, a combination with fiber-feeding is anticipated, resulting in similar problems for the H_2 method as mentioned above.

The method to detect a dependence of μ on gravitational fields, as discussed in Sect. 4, requires a high resolution satellite-based VUV spectrograph, since zero-redshift absorption by H_2 occurs in the wavelength range for which the Earth's atmosphere is opaque. Currently the COS-spectrograph aboard the HST serves this purpose although for the study of H_2 in the Milky Way it would be desirable to use a spectrometer of a resolving power beyond that of COS ($R = 20\,000$). Repair on HST instruments is likely to stop after the launch of the James Webb Space Telescope in the near future. The latter takes advantage of positioning in a Lagrange point of the Earth orbit around the sun at 1.5 million km from Earth, but is only equipped with instruments for the infrared. While in the past Galactic molecular hydrogen was first detected from a rocket-sounding mission (Carruthers 1970), thereafter dedicated VUV-spectrographs were launched in orbit: the Copernicus mission, the Orpheus telescope aboard a Space Shuttle, the Far Ultraviolet Spectroscopic Explorer (FUSE) mission, and the Space Telescope Imaging Spectrograph (STIS) and the Cosmic Origins Spectrograph (COS) instruments aboard the HST. The possibility of detecting and studying H_2 within the Milky Way, for any astrophysical purpose, is strongly endangered after HST and its instruments reach the end of their lifetime. A possible follow-up mission is that of the Large/UV/Optical/Infrared (LUVOIR) surveyor covering the H_2 absorption wavelength range, which is however still in a preparation phase (France et al. 2016).

As for the radio astronomical observations in search for a varying μ, discussed in Sect. 5, it is established that probing the methanol molecule has two distinct advantages: CH_3OH exhibits the most sensitive transitions and it exhibits transitions with a wide variety of sensitivities, such that there is no need to compare with other molecular species providing anchor lines. However, thusfar only a single absorbing system with methanol toward a radio-loud quasar is found. The object at $z = 0.88$ in the line-of-sight toward PKS1830-211 is a lensed galaxy, while the background source is a blazar and exhibiting a variable luminosity (Muller et al. 2011), which brings some difficulties when studying it in the context of μ-variation (Bagdonaite et al. 2013a). In a strategy of tightening the limitations on $\Delta\mu/\mu$ the most sensitive radio observatories could be used for finding additional methanol absorbers, preferably at higher redshifts ($z > 1$). For such survey studies the Atacama Large Millimeter Array (ALMA) observatory could be used, targeting high frequency transitions of methanol in the ALMA window (> 100 GHz). It should be noted however that the roto-torsional modes in methanol, exhibiting the largest sensitivity coefficients, are at lower frequencies, typically at < 50 GHz, depending on redshift. While currently the Effelsberg 100 m telescope (Bagdonaite et al. 2013b) and the Extended Very Large Array (Kanekar et al. 2015) are suitable observatories for the low-frequency range, in future the Square Kilometer Array (SKA) will be a very sensitive instrument to probe sensitive transitions in methanol for further constraining fundamental constants as the proton-electron mass ratio, or find a positive effect of new physics beyond the prevailing Standard Model.

Acknowledgements The author acknowledges financial support from the European Research Council (ERC) under the European Union's Horizon 2020 research and innovation programme (Grant Agreement No. 670168).

References

H. Abgrall, E. Roueff, I. Drira, Total transition probability and spontaneous radiative dissociation of B, C, B′ and D states of molecular hydrogen. Astron. Astrophys. Suppl. Ser. **141**, 297–300 (2000)

D. Albornoz Vásquez, H. Rahmani, P. Noterdaeme, P. Petitjean, R. Srianand, C. Ledoux, Molecular hydrogen in the $z_{abs} = 2.66$ damped Lyman-α absorber towards QJ0643-5041. Physical conditions and limits on the cosmological variation of the proton-to-electron mass ratio. Astron. Astrophys. **562**, 88 (2014)

J. Bagdonaite, M.T. Murphy, L. Kaper, W. Ubachs, Constraint on a variation of the proton-to-electron mass ratio from H_2 absorption towards quasar Q2348-011. Mon. Not. R. Astron. Soc. **421**, 419–425 (2012)

J. Bagdonaite, M. Daprà, P. Jansen, H.L. Bethlem, W. Ubachs, S. Muller, C. Henkel, K.M. Menten, Robust constraint on a drifting proton-to-electron mass ratio at $z = 0.89$ from methanol observation at three radio telescopes. Phys. Rev. Lett. **111**, 231101 (2013a)

J. Bagdonaite, P. Jansen, C. Henkel, H.L. Bethlem, K.M. Menten, W. Ubachs, A stringent limit on a drifting proton-to-electron mass ratio from alcohol in the early Universe. Science **339**, 46–48 (2013b)

J. Bagdonaite, E.J. Salumbides, S.P. Preval, M.A. Barstow, J.D. Barrow, M.T. Murphy, W. Ubachs, Limits on a gravitational field dependence of the proton-electron mass ratio from H_2 in white dwarf stars. Phys. Rev. Lett. **113**, 123002 (2014a)

J. Bagdonaite, W. Ubachs, M.T. Murphy, J.B. Whitmore, Analysis of molecular hydrogen absorption toward QSO B0642-5038 for a varying proton-to-electron mass ratio. Astrophys. J. **782**, 10 (2014b)

J. Bagdonaite, W. Ubachs, M.T. Murphy, J.B. Whitmore, Constraint on a varying proton-electron mass ratio 1.5 billion years after the Big Bang. Phys. Rev. Lett. **114**, 071301 (2015)

J.N. Bahcall, M. Schmidt, Does the fine-structure constant vary with cosmic time? Phys. Rev. Lett. **19**, 1294 (1967)

D. Bailly, E.J. Salumbides, M. Vervloet, W. Ubachs, Accurate level energies in the $EF^1\Sigma_g^+$, $GK^1\Sigma_g^+$, $H^1\Sigma_g^+$, $B^1\Sigma_u^+$, $C^1\Pi_u$, $B'^1\Sigma_u^+$, $D^1\Pi_u$, $I^1\Pi_g$, $J^1\Delta_g$ states of H_2. Mol. Phys. **108**, 827–846 (2010)

S.A. Balashev, V.V. Klimenko, A.V. Ivanchik, D.A. Varshalovich, P. Petitjean, P. Noterdaeme, Molecular hydrogen absorption systems in Sloan digital sky survey. Mon. Not. R. Astron. Soc. **440**, 225–239 (2014)

J. Barrow, H. Sandvik, J. Magueijo, Behavior of varying-α cosmologies. Phys. Rev. D **65**, 063504 (2002)

X. Calmet, H. Fritzsch, The cosmological evolution of the nucleon mass and the electroweak coupling constants. Eur. Phys. J. C **24**, 639–642 (2002)

G.R. Carruthers, Rocket observation of interstellar molecular hydrogen. Astrophys. J. Lett. **161**, 81 (1970)

M. Daprà, J. Bagdonaite, M.T. Murphy, W. Ubachs, Constraint on a varying proton-to-electron mass ratio from molecular hydrogen absorption toward quasar SDSS J123714.60+064759.5. Mon. Not. R. Astron. Soc. **454**, 489–506 (2015)

M. Daprà, M.L. Niu, E.J. Salumbides, M.T. Murphy, W. Ubachs, Constraint on a cosmological variation in the proton-to-electron mass ratio from electronic CO absorption. Astrophys. J. **826**, 192 (2016)

M. Daprá, P. Noterdaeme, M. Vonk, M.T. Murphy, W. Ubachs, Analysis of carbon monoxide absorption at $z_{abs} = 2.5$ to constrain variation of the proton-to-electron mass ratio. Mon. Not. R. Astron. Soc. **467**, 3848–3855 (2017a)

M. Daprá, M. van der Laan, M.T. Murphy, W. Ubachs, Constraint on a varying proton-to-electron mass ratio from H_2 and HD absorption at $z_{abs} = 2.34$. Mon. Not. R. Astron. Soc. **465**, 4057 (2017b)

J. Darling, A laboratory for constraining cosmic evolution of the fine-structure constant: Conjugate 18 centimeter OH lines toward PKS 1413+135 at $z = 0.24671$. Astrophys. J. **612**, 58 (2004)

V. Dumont, J.K. Webb, Modelling long-range wavelength distortions in quasar absorption echelle spectra. Mon. Not. R. Astron. Soc. **468**, 1568–1574 (2017)

V.A. Dzuba, V.V. Flambaum, J.K. Webb, Space-time variation of physical constants and relativistic corrections in atoms. Phys. Rev. Lett. **82**, 888–891 (1999)

V.V. Flambaum, M.G. Kozlov, Limit on the cosmological variation of m_p/m_e from the inversion spectrum of ammonia. Phys. Rev. Lett. **98**, 240801 (2007)

V.V. Flambaum, E.V. Shuryak, How changing physical constants and violation of local position invariance may occur?, in *Nuclei and Mesoscopic Physics*, ed. by P. Danielewicz, P. Piecuch, V. Zelevinsky. AIP Conference Series, vol. 995 (2008), pp. 1–11

K.C. France, B.T. Fleming, K. Hoadley, CHISL: The combined high-resolution and imaging spectrograph for the LUVOIR surveyor. J. Astron. Telesc. Instrum. Syst. **2**, 041203 (2016)

K. Griest, J.B. Whitmore, A.M. Wolfe, J.X. Prochaska, J.C. Howk, G.W. Marcy, Wavelength accuracy of the Keck HIRES spectrograph and measuring changes in the fine structure constant. Astrophys. J. **708**, 158–170 (2010)

C. Henkel, K.M. Menten, M.T. Murphy, N. Jethava, V.V. Flambaum, J.A. Braatz, S. Muller, J. Ott, R.Q. Mao, The density, the cosmic microwave background, and the proton-to-electron mass ratio in a cloud at redshift 0.9. Astron. Astrophys. **500**, 725–734 (2009)

A. Ivanchik, P. Petitjean, D. Varshalovich, B. Aracil, R. Srianand, H. Chand, C. Ledoux, P. Boissé, A new constraint on the time dependence of the proton-to-electron mass ratio. Analysis of the Q0347-383 and Q0405-443 spectra. Astron. Astrophys. **440**, 45–52 (2005)

A.V. Ivanchik, P. Petitjean, S.A. Balashev, R. Srianand, D.A. Varshalovich, C. Ledoux, P. Noterdaeme, HD molecules at high redshift: The absorption system at $z = 2.3377$ towards Q1232+082. Mon. Not. R. Astron. Soc. **404**, 1583–1590 (2010)

T.I. Ivanov, M. Roudjane, M.O. Vieitez, C.A. de Lange, W.-Ü.L. Tchang-Brillet, W. Ubachs, HD as a probe for detecting mass variation on a cosmological time scale. Phys. Rev. Lett. **100**, 093007 (2008)

P. Jansen, L.-H. Xu, I. Kleiner, W. Ubachs, H.L. Bethlem, Methanol as a sensitive probe for spatial and temporal variations of the proton-to-electron mass ratio. Phys. Rev. Lett. **106**, 100801 (2011)

P. Jansen, H.L. Bethlem, W. Ubachs, Perspective: Tipping the scales: Search for drifting constants from molecular spectra. J. Chem. Phys. **140**, 010901 (2014)

N. Kanekar, Constraining changes in the proton-electron mass ratio with inversion and rotational lines. Astrophys. J. Lett. **728**, 12 (2011)

N. Kanekar, J.N. Chengalur, Molecular gas at intermediate redshifts. Astron. Astrophys. **381**, 73–76 (2002)

N. Kanekar, C.L. Carilli, G.I. Langston, G. Rocha, F. Combes, R. Subrahmanyan, J.T. Stocke, K.M. Menten, F.H. Briggs, T. Wiklind, Constraints on changes in fundamental constants from a cosmologically distant OH absorber or emitter. Phys. Rev. Lett. **95**, 261301 (2005)

N. Kanekar, W. Ubachs, K.M. Menten, J. Bagdonaite, A. Brunthaler, C. Henkel, S. Muller, H.L. Bethlem, M. Daprà, Constraints on changes in the proton-electron mass ratio using methanol lines. Mon. Not. R. Astron. Soc. **448**, 104–108 (2015)

J. Khoury, A. Weltman, Chameleon fields: Awaiting surprises for tests of gravity in space. Phys. Rev. Lett. **93**, 171104 (2004)

J.A. King, J.K. Webb, M.T. Murphy, R.F. Carswell, Stringent null constraint on cosmological evolution of the proton-to-electron mass ratio. Phys. Rev. Lett. **101**, 251304 (2008)

J.A. King, M.T. Murphy, W. Ubachs, J.K. Webb, New constraint on cosmological variation of the proton-to-electron mass ratio from Q0528-250. Mon. Not. R. Astron. Soc. **417**, 3010–3024 (2011)

M.G. Kozlov, S.A. Levshakov, Microwave and submillimeter molecular transitions and their dependence on fundamental constants. Ann. Phys. **525**, 452–471 (2013)

P. Langacker, G. Segré, M.J. Strassler, Implications of gauge unification for time variation of the fine structure constant. Phys. Lett. B **528**, 121–128 (2002)

S.A. Levshakov, M.G. Kozlov, D. Reimers, Methanol as a tracer of fundamental constants. Astrophys. J. **738**, 26 (2011)

S.A. Levshakov, F. Combes, F. Boone, I.I. Agafonova, D. Reimers, M.G. Kozlov, An upper limit to the variation in the fundamental constants at redshift $z = 5.2$. Astron. Astrophys. **540**, 9 (2012)

A.L. Malec, R. Buning, M.T. Murphy, N. Milutinovic, S.L. Ellison, J.X. Prochaska, L. Kaper, J. Tumlinson, R.F. Carswell, W. Ubachs, Keck telescope constraint on cosmological variation of the proton-to-electron mass ratio. Mon. Not. R. Astron. Soc. **403**, 1541–1555 (2010)

V.V. Meshkov, A.V. Stolyarov, A.V. Ivanchik, D.A. Varshalovich, Ab initio nonadiabatic calculation of the sensitivity coefficients for the $X^1\Sigma_g^+ \rightarrow B^1\Sigma_u^+; C^1\Pi_u$ lines of H_2 to the proton-to-electron mass ratio. JETP Lett. **83**, 303–307 (2006)

P. Molaro, S.A. Levshakov, S. Monai, M. Centurión, P. Bonifacio, S. D'Odorico, L. Monaco, UVES radial velocity accuracy from asteroid observations. I. Implications for fine structure constant variability. Astron. Astrophys. **481**, 559–569 (2008)

S. Muller, A. Beelen, M. Guélin, S. Aalto, J.H. Black, F. Combes, S.J. Curran, P. Theule, S.N. Longmore, Molecules at $z = 0.89$. A 4-mm-rest-frame absorption-line survey toward PKS 1830-211. Astron. Astrophys. **535**, 103 (2011)

M.T. Murphy, V.V. Flambaum, S. Muller, C. Henkel, Strong limit on a variable proton-to-electron mass ratio from molecules in the distant universe. Science **320**, 1611–1613 (2008)

P. Noterdaeme, C. Ledoux, P. Petitjean, R. Srianand, Molecular hydrogen in high-redshift damped Lyman-α systems: The VLT/UVES database. Astron. Astrophys. **481**, 327–336 (2008)

P. Noterdaeme, P. Petitjean, C. Ledoux, S. López, R. Srianand, S.D. Vergani, A translucent interstellar cloud at $z = 2.69$. CO, H_2, and HD in the line-of-sight to SDSS J123714.60+064759.5. Astron. Astrophys. **523**, 80 (2010)

P. Noterdaeme, P. Petitjean, R. Srianand, C. Ledoux, S. López, The evolution of the cosmic microwave background temperature. Measurements of T_{CMB} at high redshift from carbon monoxide excitation. Astron. Astrophys. **526**, 7 (2011)

P. Noterdaeme, J.-K. Krogager, S. Balashev, J. Ge, N. Gupta, T. Krühler, C. Ledoux, M.T. Murphy, I. Pâris, P. Petitjean, H. Rahmani, R. Srianand, W. Ubachs, Discovery of a Perseus-like cloud in the early Universe—HI-to-H_2 transition, carbon monoxide and small dust grains at $z_{abs} \approx 2.53$ towards the quasar J0000+0048. Astron. Astrophys. **597**, 82 (2017)

P. Petitjean, R. Srianand, C. Ledoux, Molecular hydrogen at $z_{abs} = 1.973$ toward Q0013-004: Dust depletion pattern in damped Lyman α systems. Mon. Not. R. Astron. Soc. **332**, 383–391 (2002)

J. Philip, J.P. Sprengers, T. Pielage, C.A. de Lange, W. Ubachs, E. Reinhold, Highly accurate transition frequencies in the H_2 Lyman and Werner absorption bands. Can. J. Chem. **82**, 713–722 (2004)

H. Rahmani, M. Wendt, R. Srianand, P. Noterdaeme, P. Petitjean, P. Molaro, J.B. Whitmore, M.T. Murphy, M. Centurion, H. Fathivavsari, S. D'Odorico, T.M. Evans, S.A. Levshakov, S. Lopez, C.J.A.P. Martins, D. Reimers, G. Vladilo, The UVES large program for testing fundamental physics—II. Constraints on a change in μ towards quasar HE 0027-1836. Mon. Not. R. Astron. Soc. **435**, 861–878 (2013)

M. Rauch, The Lyman alpha forest in the spectra of quasistellar objects. Annu. Rev. Astron. Astrophys. **36**, 267–316 (1998)

E. Reinhold, R. Buning, U. Hollenstein, A. Ivanchik, P. Petitjean, W. Ubachs, Indication of a cosmological variation of the proton-electron mass ratio based on laboratory measurement and reanalysis of H_2 spectra. Phys. Rev. Lett. **96**, 151101 (2006)

E.J. Salumbides, M.L. Niu, J. Bagdonaite, N. de Oliveira, D. Joyeux, L. Nahon, W. Ubachs, CO A-X system for constraining cosmological drift of the proton–electron mass ratio. Phys. Rev. A **86**, 022510 (2012)

E.J. Salumbides, J. Bagdonaite, H. Abgrall, E. Roueff, W. Ubachs, H_2 Lyman and Werner band lines and their sensitivity for a variation of the proton-electron mass ratio in the gravitational potential of white dwarfs. Mon. Not. R. Astron. Soc. **450**, 1237–1245 (2015)

M.P. Savedoff, Physical constants in extra-galactic nebulæ. Nature **178**, 688–689 (1956)

R. Srianand, P. Petitjean, C. Ledoux, G. Ferland, G. Shaw, The VLT-UVES survey for molecular hydrogen in high-redshift damped Lyman α systems: Physical conditions in the neutral gas. Mon. Not. R. Astron. Soc. **362**, 549–568 (2005)

R.I. Thompson, J. Bechtold, J.H. Black, D. Eisenstein, X. Fan, R.C. Kennicutt, C. Martins, J.X. Prochaska, Y.L. Shirley, An observational determination of the proton-to-electron mass ratio in the early universe. Astrophys. J. **703**, 1648–1662 (2009)

W. Ubachs, E. Reinhold, Highly accurate H_2 Lyman and Werner band laboratory measurements and an improved constraint on a cosmological variation of the proton-to-electron mass ratio. Phys. Rev. Lett. **92**, 101302 (2004)

W. Ubachs, R. Buning, K.S.E. Eikema, E. Reinhold, On a possible variation of the proton-to-electron mass ratio: H_2 spectra in the line of sight of high-redshift quasars and in the laboratory. J. Mol. Spectrosc. **241**, 155–179 (2007)

W. Ubachs, J. Bagdonaite, E.J. Salumbides, M.T. Murphy, L. Kaper, Colloquium: Search for a drifting proton-electron mass ratio from H_2. Rev. Mod. Phys. **88**, 021003 (2016)

J.-P. Uzan, Varying constants, gravitation and cosmology. Living Rev. Relativ. **14**, 2 (2011)

F. van Weerdenburg, M.T. Murphy, A.L. Malec, L. Kaper, W. Ubachs, First constraint on cosmological variation of the proton-to-electron mass ratio from two independent telescopes. Phys. Rev. Lett. **106**, 180802 (2011)

D.A. Varshalovich, S.A. Levshakov, On a time dependence of physical constants JETP Lett. **58**, 237–240 (1993)

D.A. Varshalovich, A.V. Ivanchik, P. Petitjean, R. Srianand, C. Ledoux, HD molecular lines in an absorption system at redshift $z = 2.3377$. Astron. Lett. **27**, 683–685 (2001)

J.K. Webb, M.T. Murphy, V.V. Flambaum, V.A. Dzuba, J.D. Barrow, C.W. Churchill, J.X. Prochaska, A.M. Wolfe, Further evidence for cosmological evolution of the fine structure constant. Phys. Rev. Lett. **87**, 091301 (2001)

J.K. Webb, J.A. King, M.T. Murphy, V.V. Flambaum, R.F. Carswell, M.B. Bainbridge, Indications of a spatial variation of the fine structure constant. Phys. Rev. Lett. **107**, 191101 (2011)

M. Wendt, P. Molaro, QSO 0347-383 and the invariance of m_p/m_e in the course of cosmic time. Astron. Astrophys. **541**, 69 (2012)

J.B. Whitmore, M.T. Murphy, Impact of instrumental systematic errors on fine-structure constant measurements with quasar spectra. Mon. Not. R. Astron. Soc. **447**, 446–462 (2015)

J.B. Whitmore, M.T. Murphy, K. Griest, Wavelength calibration of the VLT-UVES spectrograph. Astrophys. J. **723**, 89–99 (2010)

T. Wilken, G. Lo Curto, R.A. Probst, T. Steinmetz, A. Manescau, L. Pasquini, J.I. Gonzalez Hernandez, R. Rebolo, T.W. Hänsch, T. Udem, R. Holzwarth, A spectrograph for exoplanet observations calibrated at the centimetre-per-second level. Nature **485**, 611–614 (2012)

S. Xu, M. Jura, D. Koester, B. Klein, B. Zuckerman, Discovery of molecular hydrogen in white dwarf atmospheres. Astrophys. J. Lett. **766**, 18 (2013)

Space Sci Rev (2018) 214:5
DOI 10.1007/s11214-017-0431-z

High Performance Clocks and Gravity Field Determination

J. Müller[1] · D. Dirkx[2,3] · S.M. Kopeikin[4,5] · G. Lion[6] · I. Panet[7] · G. Petit[8] · P.N.A.M. Visser[2]

Received: 12 February 2017 / Accepted: 12 October 2017 / Published online: 30 November 2017

Abstract Time measured by an ideal clock crucially depends on the gravitational potential and velocity of the clock according to general relativity. Technological advances in manufacturing high-precision atomic clocks have rapidly improved their accuracy and stability over the last decade that approached the level of 10^{-18}. This notable achievement along with the direct sensitivity of clocks to the strength of the gravitational field make them practically important for various geodetic applications that are addressed in the present paper.

Based on a fully relativistic description of the background gravitational physics, we discuss the impact of those highly-precise clocks on the realization of reference frames and

High Performance Clocks with Special Emphasis on Geodesy and Geophysics and Applications to Other Bodies of the Solar System
Edited by Rafael Rodrigo, Véronique Dehant, Leonid Gurvits, Michael Kramer, Ryan Park and Peter Wolf

The research of J. Müller has been supported by the Collaborative Research Center SFB 1128 "Relativistic Geodesy and Gravimetry with Quantum Sensors" at Leibniz Universität Hannover. The work of S.M. Kopeikin has been supported by the grant No. 14-27-00068 of the Russian Science Foundation (RSF). We gratefully acknowledge financial support from Labex FIRST-TF, ERC AdOC (grant No. 617553) and EMRP ITOC (EMRP is jointly funded by the EMRP participating countries within EURAMET and the European Union).

✉ J. Müller
mueller@ife.uni-hannover.de

D. Dirkx
D.Dirkx@tudelft.nl

S.M. Kopeikin
kopeikins@missouri.edu

G. Lion
Guillaume.Lion@obspm.fr

I. Panet
isabelle.panet@ensg.eu

G. Petit
gpetit@bipm.org

P.N.A.M. Visser
P.N.A.M.Visser@tudelft.nl

time scales used in geodesy. We discuss the current definitions of basic geodetic concepts and come to the conclusion that the advances in clocks and other metrological technologies will soon require the re-definition of time scales or, at least, clarification to ensure their continuity and consistent use in practice.

The relative frequency shift between two clocks is directly related to the difference in the values of the gravity potential at the points of clock's localization. According to general relativity the relative accuracy of clocks in 10^{-18} is equivalent to measuring the gravitational red shift effect between two clocks with the height difference amounting to 1 cm. This makes the clocks an indispensable tool in high-precision geodesy in addition to laser ranging and space geodetic techniques.

We show how clock measurements can provide geopotential numbers for the realization of gravity-field-related height systems and can resolve discrepancies in classically-determined height systems as well as between national height systems. Another application of clocks is the direct use of observed potential differences for the improved recovery of regional gravity field solutions. Finally, clock measurements for space-borne gravimetry are analyzed along with closely-related deficiencies of this method like an extra-ordinary knowledge of the spacecraft velocity, etc. For all these applications besides the near-future prospects, we also discuss the challenges that are related to using those novel clock data in geodesy.

Keywords General relativity · Relativistic geodesy · Reference frames · Time scales · High-precision time measurements · Height systems · Gravity field recovery

1 Introduction

The development of ultra-precise clocks opens the possibility for enhancing several geodetic applications, including an improved definition of timescales, datum connecting and the observation of Earth's gravity field. This paper addresses the possible impact of high-precision clocks on these applications.

The definition of timescales has wide-ranging consequences for several applications. For example, there are numerous questions associated with realizing a global reference *closest to the mean sea level*, see e.g. a discussion in Sánchez (2012). Work is still under way in

1 Institut für Erdmessung, Leibniz Universität Hannover, Schneiderberg 50, 30167 Hannover, Germany

2 Faculty of Aerospace Engineering, Delft University of Technology, Kluyverweg 1, 2629 HS, Delft, The Netherlands

3 Joint Institute for VLBI ERIC, PO Box 2, 7990 AA Dwingeloo, The Netherlands

4 Department of Physics and Astronomy, University of Missouri, 223 Physics Bldg., Columbia, MO 65211-7010, USA

5 Siberian State University of Geosystems and Technologies, Plakhotny Street 10, Novosibirsk 630108, Russia

6 LNE-SYRTE, Observatoire de Paris, PSL Research University, CNRS, Sorbonne Universités, UPMC Univ. Paris 06, LNE, 61 avenue de l'Observatoire, 75014 Paris, France

7 LASTIG LAREG, IGN, ENSG, Univ. Paris Diderot, Sorbonne Paris Cité, 35 rue Hélène Brion, 75013 Paris, France

8 Bureau International des Poids et Mesures, Pavillon de Breteuil, 92312 Sèvres Cedex, France

working groups of the International Association of Geodesy (IAG) to provide conventional definitions and procedures for the realization of a Vertical Datum Standardization.

A major objective of geodesy is the determination of physical heights, i.e. those heights related to the gravity field. The use of ultra-precise clocks to this aim has been addressed in Mai and Müller (2013), indicating that clocks with an accuracy of 10^{-18} allow for example datum connections at a level of 1 cm if combined with precise positioning by Global Navigation Satellite Systems (GNSS). Combination of high-precision terrestrial clock and GNSS data with, e.g., space-borne gravimetric data might enhance spatial resolution and accuracy of existing global gravity field models and the associated definition of vertical datums.

Finally, this paper will address the possible use of high-precision space-borne clocks for global gravimetry. This could be realized with clocks that have an ultra-high stability at short integration times (Mayrhofer and Pail 2012). An assessment will be made of the achievable gravity field retrieval performance by an efficient error propagation tool, where the performance for ultra-precise clocks will be compared with other observation techniques including the already proved techniques of low-low Satellite-to-Satellite Traking (ll-SST) and Satellite Gravity Gradiometry (SGG).

This paper is organized as follows. After introducing the relevant background on relativistic reference systems (Sect. 2) and their relation to the commonly used International Terrestrial Reference System (ITRS) in Sect. 3, a recap is provided of the required fundamental equations of time dilation (Sect. 4). The impact of high-precision clocks on the definition of timescales will be addressed in Sect. 5. The expected benefit of using high-precision clocks for terrestrial gravimetry will be discussed in Sect. 6, including their use for datum connection and height comparison (Sect. 6.3) as well as for enhancing regional gravity field solutions (Sect. 6.4). This will be followed by an assessment of using high-precision clocks for space-borne gravimetry (Sect. 7). Finally, conclusions will be drawn in Sect. 8.

2 Relativistic Celestial Reference Systems

Current theory of relativistic reference systems in the solar system has been formulated by Kopejkin (1988) and Brumberg and Kopejkin (1989a,b) and further developed by Damour et al. (1991, 1992, 1993). It was adopted by the IAU 2000 as a standard for processing high-precise astronomical and geodetic observations (Soffel et al. 2003). This theory provides precise theoretical definitions of the barycentric celestial reference system (BCRS) and the geocentric celestial reference system (GCRS) as well as the relations between them (Kopeikin et al. 2011). The important distinction of the BCRS and GCRS from their Newtonian counterparts comes from general relativity which predicts the effects of relativistic contraction of (coordinate) spatial distances and time dilation caused by the relative orbital motion of GCRS with respect to BCRS and the presence of the gravitational field of Sun, Moon, and other planets. Because of these effects the spatial coordinates of the two reference systems experience periodic relative variations due to the Lorentz and Einstein contractions of the relative order of 10^{-8}. The basic coordinate time scales for BCRS and GCRS (called TCB and TCG respectively) are also different due to the time dilation and gravitational redshift (Soffel et al. 2003).

In what follows the small letters refer to BCRS, whereas the capital letters do to GCRS. The Greek letters α, β, ... denote spacetime indices taking values 0, 1, 2, 3 with the index 0 belonging to time coordinate. Roman letters will denote the indices taking values 1, 2, 3 corresponding to the spatial coordinates only. Bold letters denote spatial vectors,

$\boldsymbol{x} = \{x^i\} = \{x^1, x^2, x^3\}$, $\boldsymbol{X} = \{X^i\} = \{X^1, X^2, X^3\}$, etc. The symbol $\mathcal{O}(c^{-n})$ means that all residual terms of order c^{-n} are neglected.

The theory of astronomical reference systems that is outlined below, is formulated in the first post-Newtonian approximation (PNA) of Einstein's theory of gravity though it can be extended to high-order approximations if necessary. The PNA is a weak-field, slow-motion approximation with three small parameters: $\epsilon = (GM/c^2R)^{1/2}$ with M and R being a characteristic mass and size of a body; $\epsilon_\omega = \omega R/c$ with ω being a rotational angular velocity of the body, and $\epsilon_v = v/c$ where v is the orbital velocity of the body. According to the virial theorem of gravitational physics (Goldstein et al. 2002) these small parameters in the solar system are ordered as follows: $\epsilon_\omega < \epsilon_v < \epsilon$, and for simplicity one is using a common denominator of the small parameters, c^{-1}, as a book-keeping parameter for a post-Newtonian expansion of the metric tensor, although it is not dimensionless. The first post-Newtonian approximation neglects terms of the order of c^{-6} in the time-time component of the metric tensor g_{00}, terms of the order of c^{-5} in the time-space components g_{0i}, and terms of the order of c^{-4} in the space-space components g_{ij}.

In Einstein's theory of gravity the gravitational field is described by a metric tensor denoted as $g_{\alpha\beta} = g_{\alpha\beta}(x^\mu)$ in BCRS and $G_{\alpha\beta} = G_{\alpha\beta}(X^\mu)$ in CGRS, that also characterizes the geometry of spacetime. The metric tensor is found by solving the Einstein field equations which preserve their form irrespectively of the choice of coordinates. Thus, they can determine the components of the metric tensor only up to four degrees of freedom. This gauge freedom of the metric tensor components corresponds to the free choice of the coordinate system. We will use the harmonic gauge condition (Fock 1964) imposed on the metric tensor in every coordinate system. The harmonic coordinates are convenient from a mathematical point of view and are often used in other branches of gravitational physics. In particular, the harmonic coordinates are a cornerstone of the IAU 2000 conventions on relativistic theory of celestial reference coordinates and time scales (Soffel et al. 2003; Petit and Luzum 2010).

Post-Newtonian form of the metric tensor in harmonic coordinates reads (Brumberg and Kopejkin 1989a,b; Damour et al. 1991)

$$\begin{aligned} g_{00} &= -1 + \frac{2w}{c^2} - \frac{2w^2}{c^4} + \mathcal{O}(c^{-6}), & G_{00} &= -1 + \frac{2V}{c^2} - \frac{2V^2}{c^4} + \mathcal{O}(c^{-6}), \\ g_{0i} &= -\frac{4w^i}{c^3} + \mathcal{O}(c^{-5}), & G_{0i} &= -\frac{4V^i}{c^3}\mathcal{O}(c^{-5}), \\ g_{ij} &= \delta_{ij}\left(1 + \frac{2w}{c^2}\right) + \mathcal{O}(c^{-4}), & G_{ij} &= \delta_{ij}\left(1 + \frac{2V}{c^2}\right) + \mathcal{O}(c^{-4}), \end{aligned} \tag{1}$$

where the gravitational potentials are functions of time and spatial coordinates: $w \equiv w(t, \boldsymbol{x})$, $w^i \equiv w^i(t, \boldsymbol{x})$ in BCRS, and $V \equiv V(T, \boldsymbol{X})$, $V^i \equiv V^i(T, \boldsymbol{X})$ in GCRS, respectively. Notice that we have used a letter V to denote the gravitational potentials of the geocentric metric tensor $G_{\alpha\beta}$. The paper by Damour et al. (1991) used W instead of V. However, the letter W is ubiquitously used in geodesy to denote the potential of the gravity force in rotating geocentric coordinates system and is represented as an algebraic sum of the gravitational potential V and the centrifugal potential, $Z = 1/2\boldsymbol{v}^2$ where $\boldsymbol{v} = \{v^i\}$ is velocity of observer with respect to GCRS (see, for example, Eq. (51) below).

The gravitational potentials are expressed in terms of the integrals taken over three-dimensional volumes occupied by matter of the bodies comprising the solar system. We single out the gravitational potentials associated with Earth and those produced by the external bodies (Moon, Sun, planets, etc.). In the post-Newtonian approximation this decomposition reads

$$w(t,\boldsymbol{x}) = w_E(t,\boldsymbol{x}) + \bar{w}(t,\boldsymbol{x}), \tag{2}$$

$$w^i(t,\boldsymbol{x}) = w^i_E(t,\boldsymbol{x}) + \bar{w}^i(t,\boldsymbol{x}), \tag{3}$$

and

$$V(T,\boldsymbol{X}) = V_E(T,\boldsymbol{X}) + \bar{V}(T,\boldsymbol{X}), \tag{4}$$

$$V^i(T,\boldsymbol{X}) = V^i_E(T,\boldsymbol{X}) + \bar{V}^i(T,\boldsymbol{X}), \tag{5}$$

where the potentials with sub-index E belong to Earth, and those with the bar to the external bodies.

Explicit expressions of the geopotentials are given in the form of a particular solution of the Poisson equations that are volume integrals,

$$w_E(t,\boldsymbol{x}) = G\int_{\mathcal{V}_E}\frac{\sigma(t,\boldsymbol{x}')d^3x'}{|\boldsymbol{x}-\boldsymbol{x}'|} + \mathcal{O}\big(c^{-2}\big), \tag{6}$$

$$w^i_E(t,\boldsymbol{x}) = G\int_{\mathcal{V}_E}\frac{\sigma^i(t,\boldsymbol{x}')d^3x'}{|\boldsymbol{x}-\boldsymbol{x}'|} + \mathcal{O}\big(c^{-2}\big), \tag{7}$$

$$V_E(T,\boldsymbol{X}) = G\int_{\mathcal{V}_E}\frac{\Sigma(T,\boldsymbol{X}')d^3X'}{|\boldsymbol{X}-\boldsymbol{X}'|} + \mathcal{O}\big(c^{-2}\big), \tag{8}$$

$$V^i_E(T,\boldsymbol{X}) = G\int_{\mathcal{V}_E}\frac{\Sigma^i(T,\boldsymbol{X}')d^3X'}{|\boldsymbol{X}-\boldsymbol{X}'|} + \mathcal{O}\big(c^{-2}\big), \tag{9}$$

where $\sigma(t,\boldsymbol{x})$, $\Sigma(T,\boldsymbol{X})$ and $\sigma^i(t,\boldsymbol{x})$, $\Sigma^i(T,\boldsymbol{X})$ are the post-Newtonian mass and mass-current densities of Earth's matter in the BCRS and GCRS respectively. The densities are directly related to the model of the energy-momentum tensor of matter distribution inside Earth. In most of the practical applications considered below it is sufficient to assume that the mass densities are approximately equal

$$\Sigma(T,\boldsymbol{X}) = \sigma(t,\boldsymbol{x}) = \rho(t,\boldsymbol{x}) + \mathcal{O}\big(c^{-2}\big), \tag{10}$$

where $\rho(t,\boldsymbol{x})$ is the baryon rest mass density. The current densities

$$\Sigma^i(T,\boldsymbol{X}) = \Sigma(T,\boldsymbol{X})(\boldsymbol{\omega}\times\boldsymbol{X})^i, \tag{11}$$

$$\sigma^i(t,\boldsymbol{x}) = \rho(t,\boldsymbol{x})\big[v^i_E + (\boldsymbol{\omega}\times\boldsymbol{x})^i\big], \tag{12}$$

where $v^i_E = v^i_E(t)$ and $\boldsymbol{\omega} = \{\omega^i(t)\}$ are the orbital velocity of Earth and the instantaneous angular velocity of Earth's rotation, respectively, and the cross between two vectors denote the standard Euclidean cross product.

Earth's gravitational potentials V_E and V^i_E admit multipolar expansion in the exterior space (outside the Earth),

$$V_E(T,\boldsymbol{X}) = \frac{GM_E}{r} + \frac{GI^i_E X^i}{r^3} + \frac{3}{2}\frac{GI^{ij}_E X^i X^j}{r^5} + \mathcal{O}\bigg(\frac{R^3_E}{r^4}\bigg), \tag{13}$$

$$V^i_E(T,\boldsymbol{X}) = \frac{G\dot{I}^i_E}{r} + \frac{G(\boldsymbol{S}_E\times\boldsymbol{X})^i}{2r^3} + \frac{\dot{I}^{ij}_E X^j}{2r^3} + \mathcal{O}\bigg(\frac{R^3_E}{r^3}\bigg), \tag{14}$$

where $r = |\boldsymbol{X}|$ is the radial distance in GCRS, M_E is the total (relativistic) mass of Earth, $\boldsymbol{S}_E = \{S_E^i\}$ is the angular momentum (spin) of Earth, I_E^i and I_E^{ij} are dipole and quadrupole moments of Earth's, and the overdot denotes a time derivative with respect to T=TCG. Usual assumption is that the Earth's center of mass is located at the origin of GCRS which makes $I_E^i = 0$, and all terms depending on the dipole moment I_E^i vanish from equations. We prefer to leave the dipole term explicitly in the equations because it is used as a vector parameter in tracking down the motion of the geocenter with respect to the origin of the International Terrestrial Reference System (ITRS)—for more details see Wu et al. (2012), Kuzin et al. (2010), Rebischung et al. (2013).

Explicit expressions of the external gravitational potentials in BCRS are given in terms of the integrals performed over the volumes of the external bodies,

$$\bar{w}(t, \boldsymbol{x}) = G \sum_{A \neq E} \int_{\mathcal{V}_A} \frac{\sigma(t, \boldsymbol{x}')d^3x'}{|\boldsymbol{x} - \boldsymbol{x}'|} + \mathcal{O}(c^{-2}), \tag{15}$$

$$\bar{w}^i(t, \boldsymbol{x}) = G \sum_{A \neq E} \int_{\mathcal{V}_A} \frac{\sigma^i(t, \boldsymbol{x}')d^3x'}{|\boldsymbol{x} - \boldsymbol{x}'|} + \mathcal{O}(c^{-2}), \tag{16}$$

where $\sigma(t, \boldsymbol{x}')$ and $\sigma^i(t, \boldsymbol{x}')$ describe the distribution of mass and mass-current densities inside the volume of the external body A.

External gravitational potentials in GCRS are found as general solutions of the Laplace homogeneous equation that are given in terms of polynomials

$$\bar{V}(T, \boldsymbol{X}) = Q_i X^i + \frac{1}{2} Q_{ij} X^i X^j + \frac{1}{6} Q_{ijk} X^i X^j X^k + \mathcal{O}(X^4), \tag{17}$$

$$\bar{V}^i(T, \boldsymbol{X}) = C_{ij} X^j + \frac{1}{2} C_{ijk} X^j X^k + \mathcal{O}(X^3), \tag{18}$$

where $Q_i = Q_i(T)$, $Q_{ij} = Q_{ij}(T)$ and $Q_{ijk} = Q_{ijk}(T)$ are the dipole, quadrupole and octupole moments of the tidal *gravitoelectric* field, $C_{ij} = C_{[ij]}(T)$, $C_{ijk} = C_{[ij]k}(T)$ are the quadrupole and octupole moments of the tidal *gravitomagnetic* field where the square parentheses around indices indicate the antisymmety.[1] The external multipole moments can be expressed in terms of the partial derivatives from the external gravitational potentials which are found by making use of the asymptotic matching of the metric tensor in BCRS and GCRS (Kopeikin et al. 2011)

$$Q_{ij} = \bar{w}_{,ij}(\boldsymbol{x}_E), \tag{19}$$

$$Q_{ijk} = \bar{w}_{,ijk}(\boldsymbol{x}_E), \tag{20}$$

$$C_{ijk} = \frac{4}{3}\left[v_E^{[i} \bar{w}^{,j]k}(\boldsymbol{x}_E) - \bar{w}^{[i,j]k}(\boldsymbol{x}_E) - \frac{1}{2}\delta^{k[i} \dot{\bar{w}}^{,j]}(\boldsymbol{x}_E) \right], \tag{21}$$

where the dot over a function denotes a total derivative with respect to time, and $v_E^i = dx_E^i/dt$ is the velocity of the geocenter with respect to BCRS. Notice that the derivatives from the external potentials are calculated on the world line of the geocenter, $\boldsymbol{x}_E = \boldsymbol{x}_E(t)$, so that, for example, $\bar{w}_{,ij}(\boldsymbol{x}_E) = \bar{w}_{,ij}(\boldsymbol{x})|_{x=x_E}$, etc.

Vector quantity Q_i is a small acceleration of the geocenter's world line with respect to a geodesic world line that is caused by the coupling of the dipole, I_E^i, and quadrupole, I_E^{ij},

[1] More precisely, we have $C_{[ij]} = 1/2(C_{ij} - C_{ji})$, and $C_{[ij]k} = 1/2(C_{ijk} - C_{jik})$.

moments of Earth's gravitational field to the tidal gravitational field of the external bodies (Kopeikin et al. 2011),

$$Q_i = \frac{1}{M_E}\left(\ddot{I}^i_E - Q_{ij} I^j_E - \frac{1}{2} Q_{ijk} I^{jk}_E\right). \tag{22}$$

The quadrupole moment, C_{ik}, is a post-Newtonian matrix of the dynamic rotation of the GCRS spatial axes with respect to the BCRS ones which appears in the metric tensor because of the IAU 2000 resolution demanding to keep the spatial axes of the BCRS and GCRS aligned to make both coordinate system kinematically-nonrotating (Brumberg and Kopejkin 1989a; Damour et al. 1991). This demand, however, makes GCRS dynamically rotating with the post-Newtonian angular velocity corresponding to the infinitesimal matrix of rotation C_{ij}. With a sufficient accuracy the mathematical expression for the matrix of the post-Newtonian dynamic rotation reads (Kopeikin et al. 2011)

$$C_{ij} = -4\bar{w}^{[i,j]}(\boldsymbol{x}_E) + 3v^{[i}_E \bar{w}^{,j]}(\boldsymbol{x}_E) + v^{[i}_E Q^{j]}, \tag{23}$$

where the first term on the right-hand side is the gravitomagnetic (Schiff or Lense-Thirring) precession, the second term is the geodetic (de Sitter) precession, and the third term is the Thomas precession caused by the non-geodesic motion of Earth's geocenter with acceleration Q_i, the square brackets around a pair of indices again indicate the antisymmetrization.

The asymptotic matching technique allows us to derive the transformation law between time and spatial coordinates of BCRS and GCRS (Kopeikin et al. 2011). The post-Newtonian transformation between the spatial coordinates of GCRS and BCRS reads (Soffel et al. 2003; Petit and Luzum 2010)

$$X^i = r^i + \frac{1}{c^2}\left[\frac{1}{2} v^i_E(\boldsymbol{v}_E \cdot \boldsymbol{r}) + \bar{w}(\boldsymbol{x}_E) r^i + r^i(\boldsymbol{a}_E \cdot \boldsymbol{r}) - \frac{1}{2} a^i_E r^2\right] + \mathcal{O}\left(c^{-4}\right), \tag{24}$$

where $\boldsymbol{r} \equiv \boldsymbol{x} - \boldsymbol{x}_E(t)$, and $\boldsymbol{a}_E$ is the coordinate acceleration of the Earth's geocenter, $\boldsymbol{a}_E = d\boldsymbol{v}_E/dt = d^2\boldsymbol{x}_E/dt^2$.

Post-Newtonian transformation of time coordinates, TCG $= T$ and TCB $= t$, is more complicated. According to the IAU 2000 Resolutions it reads (Soffel et al. 2003; Petit and Luzum 2010)

$$T = t - \frac{1}{c^2}\left[A(t) + \boldsymbol{v} \cdot \boldsymbol{r}\right] + \frac{1}{c^4}\left[B(t) + B^i(t) r^i + B^{ij}(t) r^i r^j + C(t, \boldsymbol{x})\right] + \mathcal{O}\left(c^{-5}\right), \tag{25}$$

with

$$\frac{dA(t)}{dt} = \frac{1}{2}\boldsymbol{v}^2_E + \bar{w}(\boldsymbol{x}_E), \tag{26}$$

$$\frac{dB(t)}{dt} = -\frac{1}{8}\boldsymbol{v}^4_E - \frac{3}{2}\boldsymbol{v}^2_E \bar{w}(\boldsymbol{x}_E) + 4v^i_E \bar{w}^i(\boldsymbol{x}_E) + \frac{1}{2}\bar{w}^2(\boldsymbol{x}_E), \tag{27}$$

$$B^i(t) = -\frac{1}{2}\boldsymbol{v}^2_E v^i_E + 4\bar{w}^i(\boldsymbol{x}_E) - 3v^i_E \bar{w}(\boldsymbol{x}_E), \tag{28}$$

$$B^{ij}(t) = -v^{(i}_E Q^{j)} + 2\bar{w}^{(i,j)}(\boldsymbol{x}_E) - v^{(i}_E \bar{w}^{,j)}(\boldsymbol{x}_E) + \frac{1}{2}\delta^{ij}\dot{\bar{w}}(\boldsymbol{x}_E), \tag{29}$$

$$C(t, \boldsymbol{x}) = -\frac{1}{10} r^2\left(\dot{a}^i_E r^i\right). \tag{30}$$

Here, the dot stands for the total time derivative with respect to time t, i.e.,

$$\dot{\bar{w}} \equiv \bar{w}_{,t} + v^i_E \bar{w}_{,i}, \tag{31}$$

and the round parentheses denote the symmetrization with respect to two spatial indices, for example, $B^{(ij)} = 1/2(B^{ij} + B^{ji})$.

3 International Terrestrial Reference System

The science of geodesy primarily deals with observations and measurements conducted by terrestrial observers located on Earth. Therefore, it is practically convenient to work in the rotating geocentric coordinate system—the international terrestrial reference system (ITRS) which time coordinate is TCG coordinate time T (the same as the time coordinate in GCRS) and the spatial coordinates are denoted $X^i_{\mathrm{ITRS}} = (X^1_{\mathrm{ITRS}}, X^2_{\mathrm{ITRS}}, X^3_{\mathrm{ITRS}})$. Transformation from GCRS to ITRS is given by the IERS Conventions (Petit and Luzum 2010; Kopeikin et al. 2011)

$$X^i_{\mathrm{ITRS}} = \Lambda^{ij} X^j, \tag{32}$$

where $\Lambda^{ij} \equiv \Lambda^{ij}(T)$ is the orthogonal matrix of rotation depending on time T. Due to the property of the orthogonal matrices the inverse matrix Λ^{-1} of the transformation coincides with the transpose matrix $(\Lambda^{-1})^{ij} = \Lambda^{ji}$ so that the inverse transformation between GCRS and ITRS is

$$X^i = \Lambda^{ji} X^j_{\mathrm{ITRS}}. \tag{33}$$

According to the IERS theory of the Earth rotation (Petit and Luzum 2010) the matrix Λ^{ij} can be represented in two equivalent forms corresponding to the, so called, CIO-based transformation and equinox-based transformation (see Petit and Luzum 2010, Chap. 5.9 and Kopeikin et al. 2011, Equation 9.75). For analytic consideration the equinox-based transformation is more convenient and therefore used in the following. The matrix of the equinox-based transformation is represented as a product of four matrices

$$\Lambda^{ij} = W^{ki}(T) R_3^{kp}(\mathrm{GAST}) N^{pq}(T) P^{ql}(T) B^{lj}, \tag{34}$$

where the time T=TCG, W^{ki} is the matrix of the polar wobble, R_3^{kp}(GAST) is the matrix of the diurnal rotation depending on the Greenwich Astronomical Sidereal Time (GAST) that is a function of time T=TCG, N^{pq} is the matrix of nutation, P^{ql} is the matrix of precession, and the constant matrix B^{lj} describes the, so-called, frame bias.

These matrices have the following structure (Petit and Luzum 2010)

$$P^{ij} = R_3^{ik}(\chi_A) R_1^{pk}(-\omega_A) R_3^{qp}(-\psi_A) R_1^{qj}(\varepsilon_0), \tag{35}$$

$$N^{ij} = R_1^{ki}(-\varepsilon - \Delta\varepsilon) R_3^{pk}(-\Delta\psi) R_1^{pj}(\varepsilon), \tag{36}$$

$$W^{ij} = R_3^{ki}(-s') R_2^{kp}(x_{\mathrm{p}}) R_1^{pj}(y_{\mathrm{p}}), \tag{37}$$

where ε_0 and $\varepsilon = \varepsilon(T)$ are correspondingly a constant and instantaneous obliquity of the celestial equator to ecliptic, $\chi_A = \chi_A(T)$, $\omega_A = \omega_A(T)$, $\psi_A = \psi_A(T)$ are secular variations in precession, $\Delta\epsilon = \Delta\epsilon(T)$ and $\Delta\psi = \Delta\psi(T)$ are periodic variations in nutation, $x_{\mathrm{p}} = x_{\mathrm{p}}(T)$ and $y_{\mathrm{p}} = y_{\mathrm{p}}(T)$ describe the polar wobble, and $s' = s'(T)$ is the small secular variation describing the shift between the ITRS origin of longitude and the terrestrial intermediate origin (TIO) (Kopeikin et al. 2011; Petit and Luzum 2010).

The components of the rotational matrix are used to calculate the GCRS velocity of motion of a terrestrial observer (clocks) located on Earth's surface,

$$v^i_{\mathrm{GCRS}} = \epsilon^{ijk}\omega^j X^k_{\mathrm{ITRS}} + \dot{X}^i_{\mathrm{ITRS}}, \tag{38}$$

where the overdot denotes a time derivative, ϵ^{ijk} is a fully anti-symmetric symbol of Levi-Civita with $\epsilon^{123} = +1$, X^i_{ITRS} is the ITRS coordinate of the observer (clock), $\dot{X}^i_{\mathrm{ITRS}}$ is the residual velocity of the observer (clock) with respect to ITRS due to various geophysical reasons or simply because the observer is moving in a car or aircraft, and $\omega^i = \omega^i(T)$ is the instantaneous angular velocity of the rotation of the ITRS with respect to GCRS,

$$\omega^i = -\epsilon^{ijk}\Lambda^{jp}(T)\frac{d}{dT}\Lambda^{kp}(T). \tag{39}$$

In what follows it will be convenient to split the ITRS position of the observer (clock) in two components

$$X^i_{\mathrm{ITRS}} = X^i_{\mathrm{geoid}} + \int_0^H n^i(h)dh, \tag{40}$$

where $n^i(h)$ is a unit vector along the direction of the plumb line passing through the point of observation, X^i_{ITRS} is located on Earth surface, X^i_{geoid} is the point on the geoid connected by the plumb line to the position of the observer on the surface, dh is the element of length along the plumb line, and H is the orthometric height at the observer's position (Torge and Müller 2012).

Precise analytic expression for the angular velocity, ω^i, is too complicated and we develop it only up to a linear approximation with respect to the variations of the parameters entering (35)–(37). For this purpose we use the following approximations

$$\chi_A = \delta\chi_A, \tag{41}$$

$$\psi_A = \delta\psi_A, \tag{42}$$

$$\omega_A = \varepsilon_0 + \delta\omega_A, \tag{43}$$

$$\varepsilon = \varepsilon_0 + \delta\varepsilon, \tag{44}$$

$$\mathrm{GAST} = T + \Delta\psi\cos\varepsilon_0 + \delta T, \tag{45}$$

where the very last δ-terms in (41)–(45) are functions of time, T=TCG, which are changing due to the systematic (both periodic and secular) variations in the spatial orientation of the Earth's rotational axis, wobble, and tidal friction (Kopeikin et al. 2011; Petit and Luzum 2010).

Then, in the linear approximation the components of the angular velocity of Earth's rotation are

$$\omega^1 = \dot{x}_{\mathrm{p}} + \Omega y_{\mathrm{p}} - \sin(\Omega T)\frac{d}{dT}[\delta\omega_A + \Delta\varepsilon] - \cos(\Omega T)\sin\varepsilon_0\frac{d}{dT}[\delta\psi_A + \Delta\psi], \tag{46}$$

$$\omega^2 = \dot{y}_{\mathrm{p}} - \Omega x_{\mathrm{p}} + \cos(\Omega T)\frac{d}{dT}[\delta\omega_A + \Delta\varepsilon] - \sin(\Omega T)\sin\varepsilon_0\frac{d}{dT}[\delta\psi_A + \Delta\psi], \tag{47}$$

$$\omega^3 = \Omega + \frac{d}{dT}[\delta T + \delta\chi_A - \delta\psi_A\cos\varepsilon_0], \tag{48}$$

where Ω is a fixed value of the angular velocity of the Earth rotation adopted by IAU (Petit and Luzum 2010).

4 Time Dilation Fundamental Equation

We consider a network of atomic clocks located on Earth's surface at different geographic positions. Each clock moves with respect to the GCRS along world line $X^i \equiv X^i(T)$. According to general relativity each clock measures its own proper time τ which is defined by the equation $-c^2 d\tau^2 = ds^2$ where the interval ds must be calculated along the world line of the clock. In terms of the GCRS metric tensor, the interval of the proper time reads

$$d\tau^2 = -\left(G_{00} + \frac{2}{c}G_{0i}v^i + \frac{1}{c^2}G_{ij}v^i v^j\right)dT^2, \tag{49}$$

where v^i is velocity of clock with respect to GCRS, and T=TCG. In case of clocks in space v^i is the orbital velocity of the spacecraft carrying the clock. If the clock is located on Earth's surface, the GCRS velocity of the clock is given by (38).

After replacing the GCRS metric (2) in (49) and extracting the root square, we get the fundamental time delay equation

$$\frac{d\tau}{dT} = 1 - \frac{W}{c^2} + \mathcal{O}(c^{-5}), \tag{50}$$

where time-dependent function $W = W(T)$ is given by (Kopejkin 1991; Kopeikin et al. 2016)

$$W = \frac{1}{2}v^2 + V + \frac{1}{c^2}\left(\frac{1}{8}v^4 + \frac{3}{2}v^2 V - 4v^i V^i - \frac{1}{2}V^2\right). \tag{51}$$

Function W is the post-Newtonian gravity potential taken at the point of localization of the clock. Notice that it includes both the effects of the gravitational field of the Earth and the external bodies (Sun, Moon, planets) in the form of the tidal terms as shown in equation (4) of the present paper. Furthermore, it is worth emphasizing that the relativistic corrections enter the right side of (51) not only explicitly but also implicitly through the Newtonian part because the calculation of the velocity v^i and the gravitational potential V must be completed with the post-Newtonian precision. The clock's proper time can be calculated by integrating (50) along the world line of the clock,

$$\tau = \int_{T_0}^{T}\left[1 - \frac{W(T')}{c^2}\right]dT' + \mathcal{O}(c^{-5}), \tag{52}$$

where T_0 is the initial epoch of the integration.

5 Implications for the Definition of Timescales

The 13th General Conference on Weights and Measures (CGPM) decided in its Resolution 1 in 1967[2] that *the second is the duration of 9,192,631,770 periods of the radiation corresponding to the transition between two hyperfine levels of the ground state of the caesium*

[2] The CGPM reports are available at http://www.bipm.org/fr/worldwide-metrology/cgpm/resolutions.html.

133 atom. This definition of the time unit called for the adoption of a time scale built by cumulating atomic seconds and International Atomic Time TAI was defined in 1970 by the International Committee for Weights and Measures as *the time reference established by the BIH on the basis of the readings of atomic clocks operating in various establishments in accordance with the definition of the second*, a definition recognized by the 14th CGPM in its Resolution 1 in 1971. In 1980 the definition of TAI was completed by the Consultative Committee for the Definition of the Second, adding *TAI is a coordinate time scale defined in a geocentric reference frame with the SI second as realized on the rotating geoid as the scale unit*. This definition explicitly refers to TAI as a coordinate time, hence needing a relativistic approach. In 1988 the responsibility of establishing TAI was transferred to the International Bureau for Weights and Measures (BIPM) in Sèvres (France).

TAI is not disseminated directly and Coordinated Universal Time UTC, which was designed to approximate UT1 (a timescale derived from the rotation of the Earth), was chosen as the practical world time reference. Since 1972, UTC differs from TAI by an integral number of seconds, changed when necessary by insertion of a leap second, as predicted and announced by the International Earth Rotation and Reference System Service (IERS). It is not the purpose of this paper to discuss this issue; the point is just to remind that changes in the definition of timescales may have wide-ranging consequences irrespective of the practical implications.

As mentioned in previous sections, TCB and TCG are the time coordinates of the BCRS and GCRS, respectively. The 1991 Recommendation 3 of Resolution A4 of the International Astronomical Union (IAU)[3] defined the scale unit of TCB and TCG to be consistent with the SI second. This means that if readings of the proper time of an observer, expressed in SI seconds, are recomputed into TCB or TCG using the formulas from the IAU Resolutions, without any additional scaling, one gets corresponding values of TCB or TCG in the intended units. The Recommendation also defines the origin of TCB and TCG by the following relation to TAI: TCB (resp. TCG) = TAI + 32.184 s on 1977 January 1st, 0 h TAI, at the geocenter.

In the following, we assume that the temporal variations in the physical quantities (e.g. potential, equipotential surface) due to tides are taken into account through appropriate reductions so that the quantities are considered as quasi-static, i.e. constant or with a slowly varying secular change. For a clock at rest on the Earth's surface, the relation between proper time and coordinate time is given by Eq. (50) that is

$$\frac{d\tau}{d\mathrm{TCG}} = 1 - \frac{W}{c^2}, \tag{53}$$

where for practical purposes, it is sufficient to retain in W only the first two terms of the gravity potential (rotational plus gravitational) given by Eq. (51). Any time differing from TCG by a constant rate may also be chosen as a coordinate time in the geocentric system, and this is the case of Terrestrial Time TT which differs from TCG by a constant rate:

$$\frac{d\mathrm{TT}}{d\mathrm{TCG}} = 1 - L_G. \tag{54}$$

In the original definition (IAU'1991 Recommendation 4 of Resolution A4), L_G is such as *the unit of measurement of TT is chosen so that it agrees with the SI second on the geoid*, i.e. $L_G = W_0/c^2$, where W_0 is the latest estimate of the gravity potential on the geoid. By

[3]The IAU Resolutions are available at http://www.iau.org/administration/resolutions/general_assemblies/.

IAU Resolution B9 (2000) TT is explicitly re-defined with respect to TCG by turning L_G in Eq. (54) into a defining constant where $L_G = 6.969290134 \times 10^{-10}$. This constant has been computed as $L_G = W_0/c^2$ by taking for W_0 the value 62636856 m^2/s^2 recommended by Special Commission 3 of the International Association of Geodesy (IAG) in the year 1999, before the new definition. This value will be noted below $W_0(2000)$, so that

$$\frac{d\tau}{d\mathrm{TT}} = 1 - \frac{W - W_0(2000)}{c^2}. \tag{55}$$

TAI is viewed as a realization of coordinate time TT, to within the constant offset of 32.184 s and subject to uncertainties in the realization. However there is presently a conceptual difference between the two timescales because TAI is defined with respect to the rotating geoid while TT is not, we'll come back to this later on. In principle, the transformation from proper time to TAI requires the value W_0 of the gravity potential on the actual geoid. Relation (55) transforms to

$$\frac{d\tau}{d\mathrm{TAI}} = 1 - \frac{W - W_0}{c^2}. \tag{56}$$

In practice, the potential W is expanded in the Taylor series around its value on the geoid (Torge and Müller 2012)

$$W = W_0 - \bar{g}H + \mathcal{O}(H^2), \tag{57}$$

so that Eq. (56) is generally applied as

$$\frac{d\tau}{d\mathrm{TAI}} = 1 + \frac{\bar{g}H}{c^2}, \tag{58}$$

where H is the orthometric height of the clock and $\bar{g}$ the average value of the acceleration of gravity between the geoid and the clock.

Because a change in height of one meter causes a change in rate of about 1×10^{-16}, Eq. (58) can be somewhat loosely applied for present-day Caesium primary frequency standards which have an accuracy of order 1×10^{-16}, assuming that the geoid, the reference level of the height system and the height are all correctly realized or measured with an uncertainty well below one metre. This is no more the case when considering a clock accuracy of the order 1×10^{-17} and below for which an uncertainty of order 1 cm is needed. For better accuracy, with validity about 5×10^{-19}, Eq. (53) must include tidal terms $V_{\mathrm{tide}} \equiv \bar{V}$ given in (17) to obtain

$$\frac{d\tau}{d\mathrm{TCG}} = 1 - \frac{1}{c^2}\left(\frac{1}{2}v^2 + V_E + V_{\mathrm{tide}}\right) \tag{59}$$

and

$$\frac{d\tau}{d\mathrm{TT}} = 1 - \frac{1}{c^2}\left(\frac{1}{2}v^2 + V_E + V_{\mathrm{tide}} - W_0(2000)\right). \tag{60}$$

The tidal potential has been considered in full detail, for example, in Agnew (2007). In the quadrupole gravitational field approximations it reads $V_{\mathrm{tide}} \equiv V_2$, i.e., we take into account in (17) merely the term depending on the external quadrupole moment Q_{ij}. This expression is well known in scientific literature (see, for example, Agnew 2007; Torge and Müller 2012; Simon et al. 2013), and can be presented in the form

$$V_2 = V_2^{\text{zonal}} + V_2^{\text{tesseral}} + V_2^{\text{sectorial}}, \tag{61}$$

where

$$V_2^{\text{zonal}} = \sum_{A \neq E} \mathcal{D}_A \left(\frac{\bar{r}_{\mathsf{EA}}}{r_{\mathsf{EA}}}\right)^3 \left(\frac{1}{3} - \sin^2\varphi\right)\left(1 - 3\sin^2\delta_A\right), \tag{62}$$

$$V_2^{\text{tesseral}} = \sum_{A \neq E} \mathcal{D}_A \left(\frac{\bar{r}_{\mathsf{EA}}}{r_{\mathsf{EA}}}\right)^3 \sin 2\varphi \sin 2\delta_A \cos h_A, \tag{63}$$

$$V_2^{\text{sectorial}} = \sum_{A \neq E} \mathcal{D}_A \left(\frac{\bar{r}_{\mathsf{EA}}}{r_{\mathsf{EA}}}\right)^3 \cos^2\varphi \cos^2\delta_A \cos 2h_A, \tag{64}$$

where

$$\mathcal{D}_A = \frac{3}{4} G M_A \frac{r^2}{\bar{r}_{\mathsf{EA}}^3} \tag{65}$$

is the Doodson's constant for the body A, $\bar{r}_{\mathsf{EA}}$ is the reference (mean) value of the distance r_{EA} during one revolution of the body A around Earth, δ_A, h_A are respectively the declination and the hour angle of the external body A in the ITRS, and the summation goes over all the external bodies of the solar system but Earth.

Note that all developments in this section are valid to an accuracy level of order 1×10^{-18} and should be re-examined to reach an accuracy level of order 10^{-19} ($\equiv$ mm in height). Then, e.g., the full post-Newtonian gravity potential given in Eq. (51) as well as the geoid in its relativistic definition should be considered, cf. Kopeikin et al. (2011).

Consider the transformation to TCG (the case of TAI will be considered further below) with Eq. (59), one needs to determine the gravity potential at the location of the clocks. This used to be a difficult task at the 10^{-17} level worldwide (Petit and Wolf 1997), however recent global satellite geopotential models allow this determination at the level of a few parts in 10^{18} (Denker 2013). On the other hand the centrifugal potential as well as other time-varying tidal effects due to external masses (direct effect and indirect effect from Earth and ocean tides) can reach a few 10^{-17} but can be computed at the 1×10^{-18} accuracy level (Voigt et al. 2016). In addition, other non-tidal effects due to mass redistributions in the Earth, the hydrosphere or the atmosphere are to be considered at the 10^{-18} accuracy level. It is not the purpose of this paper to study these effects but, contrary to tidal effects, it is not straightforward to model them and to define a conventional reference for these effects. Such 10^{-18} accuracy in the computation of the relativistic frequency shift will be necessary for the realization of timescales when new clocks with such frequency accuracy will provide the definition of the second.

Considering that the best Caesium primary frequency standards have an accuracy of order $1 - 2 \times 10^{-16}$, the present requirements are less stringent. In past years, it was claimed an uncertainty for the relativistic frequency shift of $2 - 3 \times 10^{-17}$ for the location of the NIST in Boulder (Colorado, USA) (Pavlis and Weiss 2003) and of 1×10^{-17} for the INRIM in Torino (Italy) (Calonico et al. 2007). A review of the frequency standards contributing to TAI in 2015[4] shows that a good fraction adopts a conservative approach in evaluating the

[4]BIPM report on Time activities available at http://www.bipm.org/en/bipm/tai/annual-report.html.

relativistic frequency shift e.g. 1×10^{-16} is used for the LNE-SYRTE in Paris (France) and for the NIM in Beijing (China), 0.5×10^{-16} for the NPL in Teddington (UK) and for the VNIIFTRI in Mendeleevo (Russia), while the NIST adopts 0.3×10^{-16} as mentioned above. Some metrological centers, however, like the INRIM or the PTB in Braunschweig (Germany) report an uncertainty at or slightly below 0.1×10^{-16}.

The geoid appearing in the definition of TAI is classically defined as the level surface of the gravity potential closest to the topographic mean sea level. Therefore the value of the potential on the geoid W_0 depends on the global ocean level which changes with time due to geophysical reasons. Several authors have considered the time variation of W_0, see e.g. (Burša et al. 2007; Dayoub et al. 2012), but there is some uncertainty in what is accounted for in such a linear model. A recent estimate (Dayoub et al. 2012) over 1993–2009 is $dW_0/dt = -2.7 \times 10^{-2}\ \mathrm{m}^2\,\mathrm{s}^{-2}\,\mathrm{yr}^{-1}$, mostly driven by the sea level change of $+2.9$ mm/yr. The rate of change of the global ocean level could vary during the next decades, nevertheless, to state an order of magnitude, considering a systematic variation in the sea level of order 3 mm/yr, different definitions of a reference surface for the gravity potential could yield differences in frequency of order 3×10^{-18} in a decade. In addition, there are numerous questions associated with realizing a global reference *closest to the mean sea level*, see e.g. a discussion in (Sánchez 2012), so that work is still under way in IAG working groups to provide conventional definitions and procedures of realization for a Vertical Datum Standardization.

Therefore, there is evidence of inconsistency in the present set of concepts and/or definitions as summarized hereafter. On the geodetic side, there is inconsistency between the three propositions:

- The geoid is the equipotential surface closest to the mean sea level;
- The value of the gravity potential on the geoid is W_0;
- W_0 is a defining (fixed) constant.

On the time metrology side, there is inconsistency between the three propositions:

- TT is a coordinate time defined by IAU Recommendation B9 (2000);
- TAI is a coordinate time with the SI second as realized on the rotating geoid as the scale unit;
- TAI is a realization of TT.

The inconsistency is at a level of 10^{-17} (equivalent to 9 cm in height) when considering that such fundamental concepts should be valid for decades.

The choice of the value $W_0(2000)$ provides a formal definition of a surface where clocks run at the same rate as TT, which has been named the chronometric geoid (Wolf and Petit 1995). As the *classical geoid* remains linked to the mean sea level, these two surfaces differ and the difference evolves with time. One possible solution on the geodetic side is to update the W_0 value as necessary to reflect the actual mean sea level as well as improvements in the determination of W_0, so that W_0 is no more used as a constant but is varying with time. On the time metrology side, one solution is to explicitly define TAI as a realization of TT so that it is no more linked to the time-varying geoid. Indeed, if the adopted value of the gravity potential on the geoid W_0 is updated, this has no implication on the L_G value which remains a fixed conventional value relating TCG and TT. The W_0 value and the definition of the geoid are the responsibility of the IAG which addressed the issue in its Resolution 1 (2015) *for the definition and realization of an International Height Reference System (IHRS)*. As mentioned in Ihde et al. (2015), *For a global height reference system, any value W_0 within a range of a few decimeters can be defined as conventional without affecting the task of defining and realizing a global height reference system*. Nevertheless,

IAG Resolution 1 (2015) recommends both a new value of the gravity potential at the geoid $W_0 = 62636853.4\ \mathrm{m^2\,s^{-2}}$, here noted $W_0(2015)$, and to use this value a reference for the IHRS. Should such IHRS-compatible geopotential values become available for vertical reference points, the difference between $W_0(2015)$ and $W_0(2000)$, equivalent to 2.9×10^{-17} in rate shift, has to be noted when realizing TT from an atomic clock. The reformulation of the TAI definition as well as its procedure of realization is under the responsibility of the Consultative Committee for Time and Frequency (CCTF) which has formed a task group proposing, at its 2017 meeting, a new definition of TAI to be adopted by the CGPM in 2018. Changing the wording in the definition of TAI will be one task but it may be the easy one. Applying Eq. (60) will still be a difficult task on a global scale at the 10^{-18} level. E.g. while height networks can be directly linked when on the same continent, larger uncertainties persist in the global determination of the geoid and of vertical references from effects affecting the sea surface topography (Fu 2010).

Placing ultra-accurate clocks on Earth is inconvenient, because of the large number of small geophysical effects that have to be taken into account to compute the relativistic frequency shift. However it is also an opportunity because their accuracy gives access to study these effects. As has long been realized and dreamed of, it might be an optimal solution to have some ultra-accurate clocks in space, to be used as a reference to generate TAI and to which Earth-based clocks could be compared. Comparisons of ultra-accurate clocks could therefore help in the future to establish a worldwide vertical datum.

6 Height Systems and Terrestrial Gravimetry with Clocks

Here, we discuss a novel method of the direct use of clock measurements to derive physical heights and to resolve discrepancies in classical height systems. We address time-variable effects that have to be considered when using clock data in practice. These effects include both temporal variations of the gravity field and instabilities in the rotation of the Earth. A particular application is the direct use of observed potential differences for the improved recovery of the regional gravity field in the Massif Central region in the middle of southern France.

6.1 Height Systems and Clock Measurements

To the major objectives of geodesy belong the determination of physical heights (i.e. those heights related to the gravity field) and the determination of the corresponding height reference surface (i.e. the geoid or quasi-geoid) (Torge and Müller 2012), see Fig. 1. For example, the orthometric height is defined by

$$H = \frac{C_P}{\bar{g}}, \tag{66}$$

where C_P is the so-called geopotential number which is the difference of the gravity potential between the geoid and the surface point P, and $\bar{g}$ is mean gravity along the (curved) plumbline. A similar relation holds for normal height H_N (where the geopotential number C_P has to be divided by a mean normal gravity value $\bar{\gamma}$) and quasi-geoid. Geopotential numbers C_P are classically obtained via geometrical levelling and terrestrial gravimetry. The latter is required to correct the raw levelling results to consider the non-parallelism of the equipotential surfaces of which the geoid is just that one surface corresponding to the

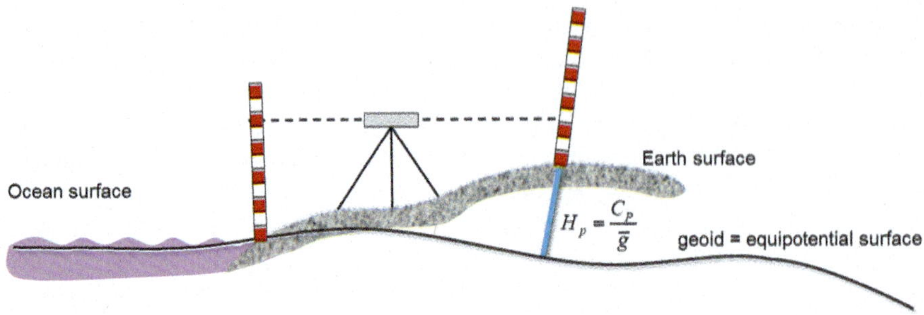

Fig. 1 Physical heights and geoid

mean ocean surface at rest. This classical approach has some drawbacks such as increasing errors with the length of the levelling loops or systematic errors by combining data from different measurement periods. This method is very time consuming, especially if large areas have to be covered. For detecting vertical deformations, repeated levelling measurements are required which further increases the mentioned complexity.

When establishing a physical height system, normally one refers to a datum point like a selected tide gauge, e.g. NAP (Normaal Amsterdams Peil), and then only differences of potential numbers ΔC_P are used further on. The basic network of physical heights for a certain area, like a country, is then computed at the level of geopotential numbers before they are converted into physical heights by dividing the C_P values by $\bar{g}$ or $\bar{\gamma}$.

If the geoid (or quasi-geoid) shall be determined point-wisely, one can use the physical heights together with ellipsoidal GNSS (Global Navigation Satellite System) heights

$$N = h - H, \tag{67}$$

where N is the geoid height above the ellipsoid, h is the ellipsoidal height of the surface point P. Similar relations hold for the quasi-geoid height: $\zeta = h - H_N$. Alternatively, one could apply the Bruns formula

$$N = \frac{T_0}{\gamma_0}, \tag{68}$$

where the geoid height N is derived from the disturbing potential T_0 at the geoid and the normal gravity value γ_0 taken at the ellipsoid. A similar expression holds again for the quasi-geoid height $\zeta = T_P/\gamma_Q$, now the disturbing potential is needed at the surface point P and the normal gravity value at the telluroid point Q.[5] The disturbing potential T_P is defined as difference between the gravity potential of the Earth W_P and the normal potential U_P which is computed from the values of a mean rotational ellipsoid approximating the Earth: $T_P = W_P - U_P$, cf. Torge and Müller (2012). The disturbing potential can be computed by solving a boundary-value problem where various variants exist. Often the long-wavelength contributions of the gravity field are taken from satellite solutions that are combined with terrestrial gravity and topographic data for the shorter wavelengths, cf. Denker (2013).

As briefly re-called from geodesy textbooks, the gravity potential and mostly differences of the gravity potential are the central quantities to derive physical heights and the geoid.

[5] The telluroid is an approximation to the Earth surface that is consistent to the theory of Molodensky (Heiskanen and Moritz 1967).

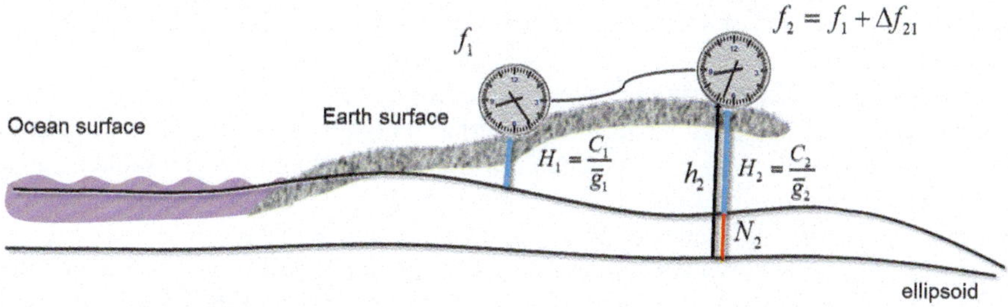

Fig. 2 Physical heights and clock measurements

In Sect. 4, the dependency of proper time to the gravity potential W is given. Now we consider the relation between physical heights and clock measurements. Two clocks in remote locations can be compared by different methods, in practice all relying on the transmission of electromagnetic signals between the two clocks (see e.g. Petit et al. 2014 and Gersl et al. 2015). We consider two clocks at locations 1 and 2 with proper times τ_1 and τ_2 (see also Fig. 2). If f_1 is the proper frequency of an electromagnetic signal measured by clock 1 at the time of emission of the electromagnetic signal, and f_2 is the proper frequency of the same electromagnetic signal measured by clock 2 at the time of reception of the signal, then

$$\frac{f_2}{f_1} = \frac{\Delta\tau_1}{\Delta\tau_2}, \tag{69}$$

where $\Delta\tau_1$ and $\Delta\tau_2$ are the increments of the proper time being reciprocal to the frequencies f_1 and f_2, respectively. By introducing Eq. (50) in Eq. (69) and assuming that $\Delta\tau = d\tau$, the corresponding proper frequency difference $\Delta f_{21} = f_2 - f_1$ is

$$\frac{\Delta f_{21}}{f_1} = \frac{\Delta W_{21}}{c^2}. \tag{70}$$

It is directly related to the difference of the gravity potential ΔW_{21} between these locations which again corresponds to the difference of the geopotential numbers $\Delta C_{21} = -\Delta W_{21}$. Thus, in future, those gravity potential values derived from clock comparisons can straightforwardly be used in physical geodesy. For example, the difference between two orthometric heights can be obtained as (see also Müller 2016)

$$\Delta H_{21} = H_2 - H_1 = H_1 \frac{\Delta\bar{g}_{12}}{\bar{g}_2} + \frac{\Delta C_{21}}{\bar{g}_2} \tag{71}$$

with the difference of the mean gravity values along the plumbline at both locations $\Delta\bar{g}_{12} = \bar{g}_1 - \bar{g}_2$ and the difference of the geopotential numbers $\Delta C_{21} = C_2 - C_1 = W_1 - W_2$, which can now be derived from the clock comparisons according to Eq. (70):

$$\Delta H_{21} = H_1 \frac{\Delta\bar{g}_{12}}{\bar{g}_2} - \frac{c^2}{\bar{g}_2}\frac{\Delta f_{21}}{f_1}. \tag{72}$$

6.2 Time-Variable Effects on Clock Measurements

In the following, we want to discuss some properties that have to be considered when using clock measurements and we will give a few examples for potential applications where clock data might especially be beneficial.

At the accuracy level of mm to cm, it is not sufficient anymore to consider gravity potential values as constant in time. They are affected by deformations (up to 40 cm due to tidal effects) and mass variations. In the following, we always indicate the combined effect for both contributions (deformation plus mass effect) as sensed by the clocks. And to make the effect more illustrative, the gravity potential value is converted into a height value by dividing it by some mean gravity value of about 9.81 $\mathrm{m/s^2}$. Even if only potential differences are considered, there are remarkable effects of up to 8 cm between PTB in Braunschweig and NPL in London caused by relative perturbations of the solid Earth tides (Voigt et al. 2015, 2016). Those variations with changing amplitudes have periods of 12 hours and longer. In addition, many smaller effects have to be considered: periodic effects due to ocean tides (a few cm), non-tidal oceanic and atmospheric effects (1 cm) or episodic effects, e.g., caused by storm surges or droughts (1 cm). Further effects comprise pole tides, variations due to land hydrology, tectonic and GIA-induced processes, etc., see Voigt et al. (2016). Another class of time-variable effects on the gravity potential are variations in Earth orientation.

The Earth orientation effects affect the rate of clocks located on the Earth surface as they are produced by the time-dependent changes in the velocity-dependent potential of the centrifugal force (Fateev et al. 2015). The centrifugal force potential consists of three terms as a direct consequence of the velocity decomposition equation (38)

$$\frac{1}{2}\boldsymbol{v}^2 = \frac{1}{2}\boldsymbol{v}^2_{\mathsf{GCRS}} = \frac{1}{2}(\boldsymbol{\omega}\times\boldsymbol{X}_{\mathsf{ITRS}})^2 + (\boldsymbol{\omega}\times\boldsymbol{X}_{\mathsf{ITRS}})\cdot\dot{\boldsymbol{X}}_{\mathsf{ITRS}} + \frac{1}{2}\dot{\boldsymbol{X}}^2_{\mathsf{ITRS}}. \tag{73}$$

The first term on the right hand side of (73) is the centrifugal potential caused by the rotation of ITRS in space. We shall denote it as

$$Z \equiv \frac{1}{2}(\boldsymbol{\omega}\times\boldsymbol{X}_{\mathsf{ITRS}})^2, \tag{74}$$

and decompose it in several components corresponding to secular changes in the rotational velocity, precesssion, nutation, and the polar wobble. It is convenient to write down the decomposition in terms of geocentric spherical coordinates associated with ITRS,

$$X^1_{\mathsf{ITRS}} = r\cos\varphi\cos\lambda, \qquad X^2_{\mathsf{ITRS}} = r\cos\varphi\sin\lambda, \qquad X^3_{\mathsf{ITRS}} = r\sin\varphi. \tag{75}$$

The centrifugal potential can be represented as a superposition

$$Z = Z_{\oplus} + Z_{\mathrm{sec}} + Z_{\mathrm{prec}} + Z_{\mathrm{nut}} + Z_{\mathrm{wob}}, \tag{76}$$

where

$$Z_{\oplus} = \frac{1}{2}\Omega^2 r^2\cos^2\varphi, \tag{77}$$

$$Z_{\mathrm{sec}} = \Omega r^2\cos^2\varphi\frac{d}{dt}[\delta T + \delta\chi_A - \delta\psi_A\cos\varepsilon_0], \tag{78}$$

$$Z_{\mathrm{prec}} = \frac{1}{2}\Omega r^2\sin 2\varphi\left[\cos(\Omega t-\lambda)\sin\varepsilon_0\frac{d\delta\psi_A}{dt} + \sin(\Omega t-\lambda)\frac{d\delta\omega_A}{dt}\right], \tag{79}$$

$$Z_{\mathrm{nut}} = \frac{1}{2}\Omega r^2\sin 2\varphi\left[\cos(\Omega t-\lambda)\sin\varepsilon_0\frac{d\Delta\psi}{dt} + \sin(\Omega t-\lambda)\frac{d\Delta\epsilon}{dt}\right], \tag{80}$$

$$Z_{\mathrm{wob}} = \frac{1}{2}\Omega^2 r^2\sin 2\varphi(x_p\sin\lambda + y_p\cos\lambda). \tag{81}$$

The second term in (73) describes the coupling of the rotational velocity of ITRS and the residual velocity $\dot{X}_{\mathrm{ITRS}}$ of clocks due to their motion with respect to ITRS. Taking the time derivative from the ITRS coordinates (75) of clocks, substituting it to the second term in the right hand side of (73) and calculating, we get

$$Z_c \equiv (\omega \times X_{\mathrm{ITRS}}) \cdot \dot{X}_{\mathrm{ITRS}} = \Omega r^2 \cos^2 \phi \dot{\lambda} = 2Z_\oplus \frac{\dot{\lambda}}{\Omega}. \tag{82}$$

We can observe that the vertical speed of clocks and the motion of the clocks along the meridian does not contribute to Z_c, only the longitudinal component of the velocity matters. The term Z_c exceeds 10^{-18} if the tangential residual velocity of clocks on the Earth's equator exceeds 0.2 mm/s. We shall assume that clocks are at rest which allows us to neglect the contribution of Z_c to the velocity-dependent potential along with the third term in (73). In important cases of clock transportations on cars or aircrafts the term Z_c must be retained. All rotational effects can well be modelled. In future, some of those time-variable effects might even be measured in clock networks.

6.3 Clocks for Datum Connection and Resolving Uncertainties in Height Systems

What is the major benefit of highly-precise clock measurements? Clocks can connect distant areas, i.e. they deliver physical height differences for the observed points without being affected by levelling errors or by some smoothing effect when combined global gravity field models, e.g., from GOCE are used. No (good) ground gravity data from terrestrial gravimetry is needed to fill-in regional to local gravity variations. This might especially be interesting in (relatively) unsurveyed countries where no gravity data are available at all, where terrestrial data are difficult to obtain (e.g. in rain forests) or in areas with rough environment (e.g. in mountain areas).

Clock measurements can be used as independent observations to resolve discrepancies in (classical) realisations of height systems and geoid solutions, see example in the Alps (Denker et al. 2015). Even in Germany between north and south over only 700 km, there is a 4 cm discrepancy when determining the quasi-geoid difference by different methods, such as GNSS/levelling versus gravimetric solutions. A similar comparison between NPL, London and PTB, Braunschweig showed differences between the classical methods of up to 15 cm (Denker 2016). These differences are most probably caused by geometric levelling that accumulates measurement errors with increasing distance.

Nevertheless in Europe, the connection of national height systems (i.e. of the specific vertical datums of each country) might be possible by applying standard geodetic methods—with the above mentioned uncertainties. But in regions like South America, large differences with large error bars are present when comparing various tide gauges, especially in the southern part (see Fig. 3 taken from Sánchez 2015) where differences in the decimeter level are present. Here, immediate improvements could be achieved, even with recent clock technology at the few cm level of accuracy, assuming that these clocks could be placed at those tide gauges and be connected by dedicated fiber links. Besides just taking raw difference measurements, one could carry out a new adjustment for the height differences where clock, GNSS and altimetry data are used as joined input.

Another application could be running clocks on ships to independently control altimetric measurements. The ocean surface roughly approximates an equipotential surface in the Earth gravity field (then clock rates would not vary), but deviate from it because of ocean currents like the gulf stream. These deviations reach up to 1 to 2 m and are observed by

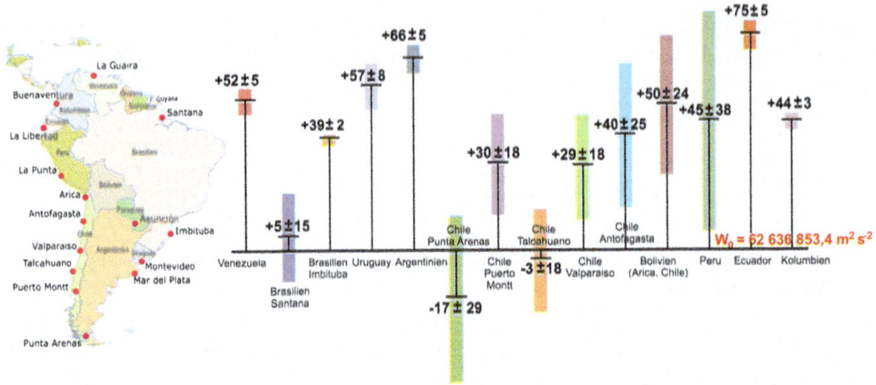

Fig. 3 Difference of regional vertical reference levels to a global one (related to W_0) in South America derived from geodetic measurements (gravity, GNSS, levelling, satellite altimetry), unit: cm (figure is taken from Sánchez 2015)

satellite altimetry together with the marine geoid, e.g., derived from GOCE data (Pail et al. 2010). One idea now is to use combined clock and GNSS measurements along selected profiles that can be compared to the results of altimetric measurements. Once again, possible systematics might be revealed by the clock measurements. Running a clock on a moving platform without interruption, however, is another challenge that is not discussed here. And, it might even be more challenging when the moving clock has to be (frequently) compared to a remote clock with the same accuracy.

6.4 Clocks for High Resolution Geopotential Recovery

To illustrate what could be the benefit of clocks for high-resolution geopotential recovery, beyond the resolution of the satellites and as complement to near-surface gravity data, we follow the work by Lion (2017) and perform simple simulations where we compare the quality of the geopotential reconstruction with and without adding clock data. We consider the Massif Central area, marked by smooth, moderate altitude mountains and volcanic plateaus, see Fig. 4 (left), which leads to variations of the gravitational field over a range of spatial scales, as illustrated in Fig. 4 (middle and right).

For the tests, we proceed as follows. We sample synthetic gravity and potential data from a spherical harmonics geopotential model, and we build a geopotential control grid. We estimate the gravity potential by least-squares collocation from the synthetic data. We compare the recovered potential on the control grid, to the reference one to assess the improvements that can be obtained using clock data. In this process, different parameters of the problem can be varied, as the spatial density of clock data for instance, studied hereafter.

In more details, the main steps of this methodology are thus:

1. High spatial resolution 5-km step grids of gravity anomalies $\overline{\delta g}$ and the disturbing potential $\overline{T}$ are generated with the program GEOPOT (Smith 1998), which allows to compute the gravity field related quantities at given locations by using mainly a geopotential model. These reference solutions, see Fig. 4, are obtained by using a state of the art geopotential model (Förste et al. 2014, EIGEN-6C4) up to degree and order 2000 (10 km resolution). The long wavelengths of the gravity field covered by the satellites and longer than the extent of the local area are removed up to 200 km resolution, providing centered

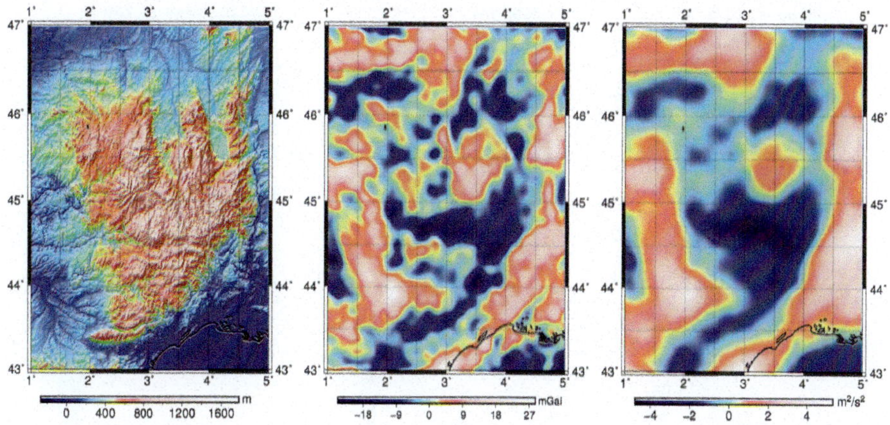

(a) Topography (b) $\overline{\delta g}$ (c) $\overline{T}$

Fig. 4 Topography and reference grids of the synthetic field $\overline{\delta g}$ and $\overline{T}$ in Massif Central area. Anomalies are computed at the Earth's surface from the EIGEN-6C4 model up to d/o 2000. Topography is obtained from the 30 m digital elevation model over France by IGN (Institut National de l'Information Géographique et Forestière), completed with Smith and Sandwell (1997) bathymetry and SRTM (Shuttle Radar Topography Mission) data

data or close to zero for the determination of local covariance function. The terrain effects are removed with the help of the topographic potential model dV_ELL_RET2012 (Claessens and Hirt 2013);
2. The distribution of gravimetric data is chosen to be the same than the control grid and contains 6989 location points. Several spatial distributions of clocks, more and less dense, are generated by randomly sampling points from this regular grid, see Fig. 5 (top); and synthetic measurements δg^* and T^* are simulated on these points as previously;
3. A white noise is added to the simulated data δg^* and T^*, with a standard deviation of $0.1\ \mathrm{m^2\,s^{-2}}$ (i.e. 1 cm on the geoid) for clocks and 1 mGal for gravimetric measurements;
4. The disturbing potential $\tilde{T}$ is estimated from the 6989 synthetic measurements δg only, and from the combination T^* and δg^* on the 5-km step grid using the Least-Square Collocation (LSC) method (Moritz 1980). In this step, we make an assumption on the gravity field regularity in the target area, using a logarithmic 3D covariance function (Forsberg 1987). This model has the advantage to provide the auto-covariances (ACF) and cross-covariances (CCF) of the potential T^* and its derivatives in closed-form expressions. Parameters of this model are adjusted on the empirical ACF of δg^* with the program GPFIT (Forsberg and Tscherning 2008). Note that this covariance function contains low frequencies that we have removed from the data in step 1;
5. Finally, the potential recovery quality is evaluated for all the data distribution sets and types of data by comparing the statistics of the residuals between $\tilde{T}$ and $\overline{T}$.

Figure 5(b) shows the residuals between the original and the reconstructed potential for denser and denser distributions of clocks. Clock number zero means that only gravity data have been used.

With no clock, the geoid is reconstructed with an accuracy of 3 cm in rms (or $0.3\ \mathrm{m^2\,s^{-2}}$ in term of potential variations). When clocks with an accuracy of 1-cm for geoid height differences (or $0.1\ \mathrm{m^2\,s^{-2}}$) are added to the network, they improve the rms by several factors and reduce the bias up to 2–3 orders of magnitude. Also, we can see that it is not necessary to have a dense clock coverage, beyond 10 percent of the clock number of gravity points,

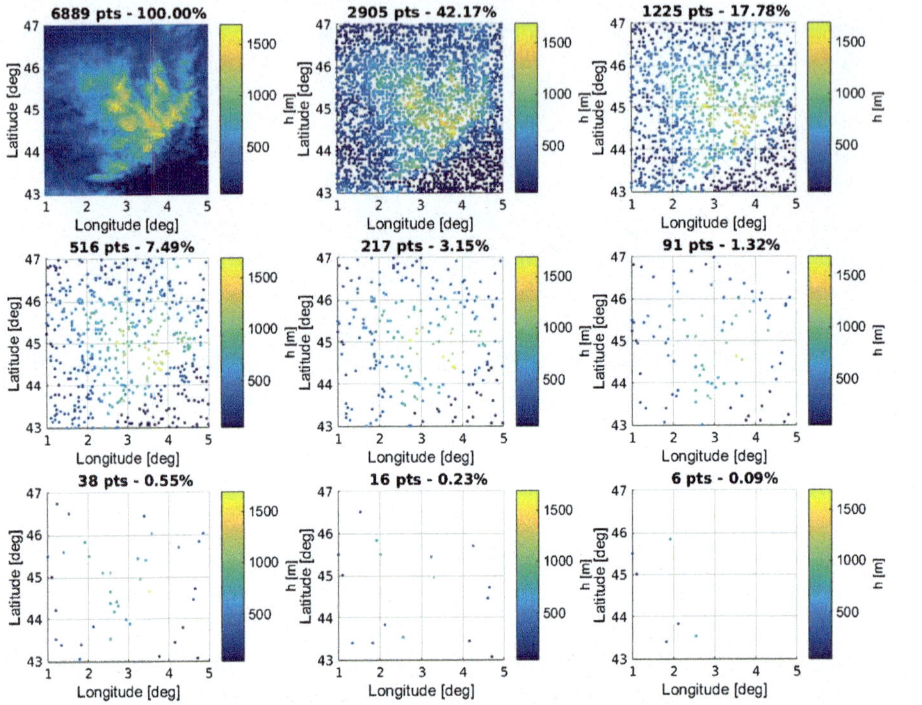

(a) Clock coverages randomly generated from a regular 5-km step grid.

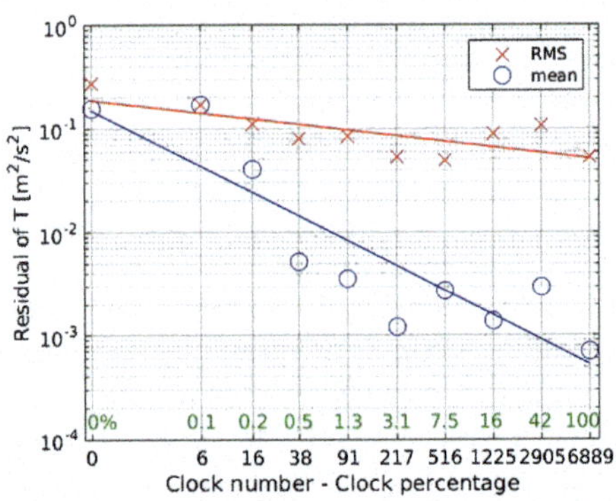

(b) Performance of the potential reconstruction (expressed by the rms and mean of differences between the original potential on the regular grid and the reconstructed one) wrt the number of clocks. 6989 gravity data have been used in this test. In green: number of clock data in terms of percentage of δg data.

Fig. 5 Contribution of clocks to the reconstruction of the disturbing potential in the Massif Central area

to improve the determination of the disturbing potential. In this test, a coverage of around 1–3 percent of the gravity data could be sufficient to reach the centimeter-level precision and improve greatly the bias.

This simple example shows that clock-based geodetic observables provide useful information at spatial scales beyond those of the satellites for geopotential determination. The gain would be even larger in areas of higher relief. A more detailed study discussing the role of different parameters, such as the noise level in the data, effects of the resolution of gravity measurements and modeling errors can be found in Lion et al. (2017), in which they consider realistic spatial samplings. The addition of gravimetric data outside the modeling target area could also be considered in future works.

7 Space-Borne Gravimetry with Clocks

Sustained observation of the time varying gravity field of the Earth with the highest possible precision and spatio-temporal resolution is crucial for understanding mass transport in the Earth system, which contributes to a better understanding of all kinds of geophysical processes and their relation to climate change, geohazards, etc. (ESA 2015). The US/German Gravity Recovery and Climate Experiment (GRACE) has shown the tremendous potential of observing mass changes due to all kinds of processes, including continental hydrology (Tapley et al. 2004), ice melt in Greenland and West-Antarctica (Velicogna and Wahr 2006a,b), and depletion of huge aquifer systems (Richey et al. 2015). The importance of sustained observation is reflected by the realization of the upcoming GRACE Follow-On mission (Sheard et al. 2012).

GRACE-type missions rely predominantly on so-called low-low satellite-to-satellite tracking (ll-SST), where the distance between two trailing satellites, typically with a distance of the order of 100–200 km, is measured with a precision of a few microns for microwave systems (Dunn et al. 2003). This technique is especially suitable for observing mass changes with a temporal resolution of the order of weeks to months at spatial resolutions of at best a few hundreds of kms. The European Space Agency (ESA) Gravity field and steady-state Ocean Circulation Explorer (GOCE) used the concept of satellite gravity gradiometry (SGG), where minute acceleration differences between adjacent capacitive accelerometers (baseline ≈ 0.5 m) are observed (Drinkwater et al. 2007). This concepts allows the observation of the gravity field with a spatial resolution of around 70–100 km, but is not well-suited for observing temporal gravity field changes.

Sustaining and at the same time improving the observation of mass transport through space-borne gravimetry has led to the selection of GRACE Follow-on. This mission will not only carry a radiowave instrument, but also a laser link for providing range observations with a precision that is claimed to be a factor 20 better than the GRACE microwave system (Jet Propulsion Laboratory 2017). In addition, studies have been carried out that demonstrate a great potential for enhancing both the temporal and spatial resolution of flying multiple ll-SST missions in parallel, and at the same time also enhancing the global isotropy and homogeneity of gravity field solutions (Thales Alenia Spazio 2010). However, there is a limit of what will be possible in terms of spatial and temporal resolution when using ll-SST or GOCE-type SGG. Therefore, new concepts are under study, including the use of cold atom interferometry (Carraz et al. 2015), and the use of ultra-precise space-borne atomic clocks (Mayrhofer and Pail 2012).

7.1 Space-Borne High-Precision Clocks

Here, we analyze the use of clock comparison between a LEO satellite and terrestrial ground stations for the purpose of determining the Earth's gravity field. The frequency of an electromagnetic signal transmitted from the ground and observed at the spacecraft will be shifted because of both the relative velocity and the difference in gravitational potential between the transmitter and receiver. By combining the measured frequencies, the following observable is obtained (Kopeikin et al. 2011):

$$\frac{f_1}{f_2} = \frac{d\tau_2}{d\tau_1} = \left(\frac{d\tau}{dT}\right)_2 \frac{dT_2}{dT_1} \left(\frac{dT}{d\tau}\right)_1, \tag{83}$$

where T and τ denote the TCG coordinate time and observer proper time, respectively. The 2 and 1 subscripts denote the receiver (spacecraft) and transmitter (ground station). The terms $d\tau/dT$ are given to post-Newtonian order by Eq. (50). The ratio of coordinate times (in TCG) can, to first order, be expressed as follows:

$$\frac{dT_2}{dT_1} = \frac{1 - \frac{\hat{\mathbf{k}}\cdot\mathbf{v}_2}{c}}{1 - \frac{\hat{\mathbf{k}}\cdot\mathbf{v}_1}{c}} \approx 1 - \frac{\hat{\mathbf{k}}\cdot\mathbf{v}_{21}}{c}, \tag{84}$$

where $\mathbf{v}_2$ and $\mathbf{v}_1$ denote the velocity vectors of receiver and transmitter (evaluated at T_2 and T_1), and $\mathbf{v}_{21} = \mathbf{v}_2 - \mathbf{v}_1$. In the above, $\hat{\mathbf{k}}$ denotes the unit vector from transmitter position $\mathbf{r}_1$ to receiver position $\mathbf{r}_2$, so:

$$\hat{\mathbf{k}} = \frac{\mathbf{r}_2(T_2) - \mathbf{r}_1(T_1)}{|\mathbf{r}_2(T_2) - \mathbf{r}_1(T_1)|}. \tag{85}$$

The relation in Eq. (84) is the so-called radial (or first-order) Doppler effect: the direct result of a change in transmission coordinate time on the reception coordinate time. In the following, we will assume that $d\tau/dT$ at the ground station can be modelled/measured to an uncertainty that is much better than for the space segment. The influence on the velocity of the spacecraft on its local $d\tau/dT$ is termed the quadratic (or second-order) Doppler effect.

For this application, we wish to retrieve the information on the Earth's gravitational potential V_E from measurements of f_2/f_1. However, the measured frequency ratio encodes both velocity and potential differences. To map the influence of both spacecraft velocity uncertainties and the strength of the potential signature to uncertainties in the model for $\Delta(f_2/f_1)$, we write for the first-order errors:

$$\Delta\left(\frac{f_2}{f_1}\right)_{\text{radial Doppler}} = \left|\frac{d}{d(\hat{\mathbf{k}}\cdot\mathbf{v}_{21})}\left(\frac{dT_2}{dT_1}\right)\right| \Delta(\hat{\mathbf{k}}\cdot\mathbf{v}_{21}) \approx \frac{\Delta(\hat{\mathbf{k}}\cdot\mathbf{v}_{21})}{c}, \tag{86}$$

$$\Delta\left(\frac{f_2}{f_1}\right)_{\text{quadratic Doppler}} = \left|\frac{d}{dv_2}\left(\frac{d\tau}{dT}\right)_2\right| \Delta(v_2) \approx \frac{v_2\Delta(v_2)}{c^2}, \tag{87}$$

$$\Delta\left(\frac{f_2}{f_1}\right)_{\text{Einstein}} = \left|\frac{d}{dV_E}\left(\frac{d\tau}{dT}\right)_2\right| \Delta(V_E) \approx \frac{\Delta(V_E)}{c^2}, \tag{88}$$

where the $\Delta(*)$ denotes the uncertainty in the parameter/measurement $*$. By combining Eq. (88), with Eqs. (86), and (87), respectively, we obtain the uncertainty in the potential determination $\Delta(V_E)$ induced by the velocity uncertainty as a result of the radial and quadratic Doppler effect, due to $\Delta(\hat{\mathbf{k}}\cdot\mathbf{v}_{21})$ and $\Delta(v_2)$, respectively.

The influence of the radial Doppler effect is by far the greatest. The maximum allowable error in $\Delta(\hat{\mathbf{k}}\cdot\mathbf{v}_{21})$ needed to obtain a given uncertainty ΔV_E for the potential is obtained from Eqs. (86) and (88):

$$\Delta(\hat{\mathbf{k}}\cdot\mathbf{v}_{21}) = \frac{\Delta V_E}{c}. \tag{89}$$

A potential height error Δh_g corresponds to a potential error of $\Delta h_g \cdot g$, with g the gravitational acceleration. Consequently, for Δh_g=1 cm, we get $\Delta(\hat{\mathbf{k}} \cdot \mathbf{v}_{21}) = \frac{0.01g}{c} \approx 3 \times 10^{-10}$ m/s or 0.3 nm/s (assuming a LEO satellite flying at 7.5 km/s with $g \approx 9$ m/s^2). This value is well below the orbit uncertainty of state-of-the-art orbit determination. However, it is possible to largely eliminate the radial Doppler shift by using a coherent two-way optical link (Djerroud et al. 2010; Blanchet et al. 2001). It is not possible, however, to eliminate the influence of the quadratic Doppler effect, since we cannot separate the influence of v and V_E in the equation for $(d\tau/dT)_2$ by combining a one- and two-way link. Again assuming an influence of maximum 1 cm on the geoid determination, the quadratic Doppler term results in a required spacecraft velocity uncertainty of better than $\frac{0.01 \times g}{7.5 \times 10^3} \approx 12$ μm/s.

In light of the influence of the radial Doppler effect, it will be crucial to have a two-way link, in addition to the one-way frequency comparison. In the error simulations below, it is therefore assumed that such a two-way link is available.

7.2 Accuracy Assessment

A hypothetical mission is defined where a satellite is flying in a low Earth orbit carrying an ultra-precise clock that can be compared continuously with perfect reference clocks. It is assumed that this satellite flies in a circular orbit, where its projection on the surface of the Earth (the so-called ground track) repeats after a certain number of nodal days. For such an observation geometry, use can be made of an efficient error propagation method that was introduced in Colombo (1984). This error propagation tool has been compared and validated by comparison with robust numerical integration methods and end-to-end simulations for several observation types, including orbit perturbations, SGG and ll-SST (Schrama 1991; Visser et al. 2001).

The satellite is assumed to fly in a polar orbit (inclination $i = 90°$) for which the ground track repeats every month (31 days) in which the satellite completes 497 orbital revolutions. The height above the Earth's surface of the satellite is about 250 km. The space-borne clock is assumed to have a frequency dependent instability equal to $30 \times 10^{-18}/\sqrt{\tau}$, where τ represents the clock integration time. This instability is about equivalent to a stability of 10^{18} for an integration time τ of 1000 s, cf. Giorgetta et al. (2013). Assuming the clock instability to be the only source of measurement error $\Delta(f_2/f_1)$, Eq. (88) leads to the following error for the observation of the gravity potential:

$$\frac{\Delta V_{E,\sigma(clock)}}{g} \approx 30 \text{ cm}/\sqrt{\tau}. \tag{90}$$

According to Eq. (87), the velocity then has to be known with a precision of about:

$$\Delta v = 360 \text{ μm/s}/\sqrt{\tau}. \tag{91}$$

Determining the LEO satellite velocity with such a precision is very challenging, but might be feasible with future optical links (Chiodo et al. 2013). The requirement for the velocity determination is based on the assumption that the velocity uncertainty causes an error in the potential that is equal or smaller than the measurement error due to clock instability.

The gravitational potential is expanded in terms of spherical harmonics. The uncertainty in the gravity field coefficients is then written as follows, where we neglect the uncertainty in r, λ and ϕ:

$$\frac{\Delta V_E}{g} = \frac{1}{g}\frac{\mu}{r}\left\{1 + \sum_{l=2}^{\infty}\sum_{m=0}^{l}\left(\frac{a_e}{r}\right)^l (\Delta\bar{C}_{lm}\cos m\lambda + \Delta\bar{S}_{lm}\sin m\lambda)\bar{P}_{lm}(\sin\phi)\right\}, \tag{92}$$

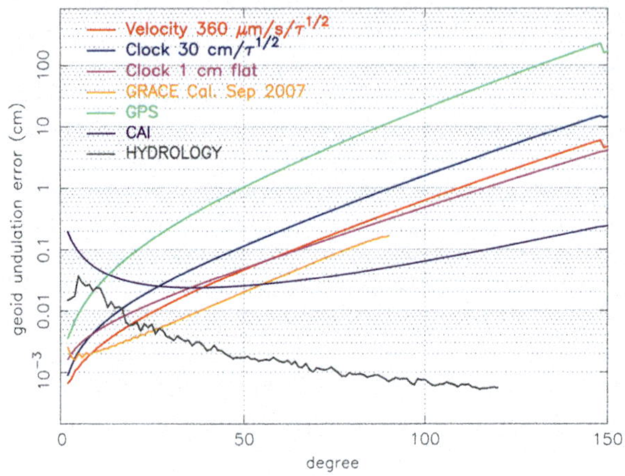

Fig. 6 Predicted gravity field retrieval errors for a LEO satellite in a 250-km altitude circular orbit with monthly repeating ground track. The satellite is equipped with a GPS receiver allowing 1-cm precision orbit determination (GPS), a 3-dimensional gradiometer based on Cold Atom Interferometry with flat noise of 0.4 mE (CAI), an ultra-precise clock (clock 30 $\mathrm{cm}/\tau^{1/2}$), instrumentation for determining the 3-dimensional velocity (velocity 360 $\mu\mathrm{m/s}/\tau^{1/2}$), and an ultra-precise clock resulting in 1 cm potential height flat noise (clock 1 cm flat). A typical curve for the claimed GRACE performance (GRACE Cal Sep. 2007), and based on a typical monthly gravity change due to continental hydrological mass transport (HYDROLOGY) are included for reference

where μ is the gravity parameter (the product of the universal gravitational constant G and Earth's mass M), a_e is the mean equatorial radius, r, ϕ, λ are the spherical coordinates (radius, geocentric latitude, longitude) denoting the location of the space-borne clock, $\bar{P}_{lm}$ is the normalized Legendre polynomial of spherical harmonic degree l and order m, and $\Delta\bar{C}_{lm}$, $\Delta\bar{S}_{lm}$ represent the estimated gravity field coefficients.

For a circular satellite orbit with repeating ground track, Eq. (92) can be represented by a Fourier series (Visser et al. 2003) and the method outlined in Colombo (1984) can be used for predicting the gravity field retrieval performance. The gravity field can be determined from time series of observed or determined satellite velocities as well. In that case, use is made of a linear orbit perturbation theory, where the associated equations can also be represented by a Fourier series (Schrama 1991). Then also the method in Colombo (1984) can be used for a gravity field retrieval accuracy assessment. Please note that in case of deriving gravity from satellite velocity perturbations, it is assumed that non-gravitational accelerations are known through e.g. observation by precise accelerometers or absent due to a drag-free control system.

The predicted gravity field retrieval performances according to the error spectra in Eqs. (90) and (91) are included in Fig. 6. It can be observed that the very precise knowledge of the satellite velocity leads to better gravity field retrieval performance through classical orbit perturbation analysis as compared to the post-Newtonian derivation of potential height from clock observations. For comparison, the predicted gravity field performances for the same LEO satellite are included if equipped with a high-quality Global Navigation Satellite System (GNSS) receiver allowing observation of orbit position perturbations with 1 cm precision, and a gradiometer based on cold atom interferometry (0.4 mE flat noise). We note that this assumption on the uncertainty is optimistic and does not reflect the state of the art for a space instrument, which is about one order of magnitude worse. With the used assumptions, the ultra-stable clock does show a much better gravity field performance than the GPS receiver, and also the gradiometer based on cold atom inferometry up to spherical harmonic degree 30. Figure 6 also includes a performance prediction for a clock that would have a flat error spectrum with a stability of 10^{18} leading to 1 cm error in potential height. Such a—highly speculative—performance would allow the observation of time variable gravity with a precision that is relatively close to the currently claimed performance level for GRACE Release 5 Level-2 products (Dahle et al. 2013, see also Fig. 6).

Being able to determine the 3-dimensional velocity of the LEO satellite with a precision according to Eq. (91) leads to a performance that is starting to become competitive with the currently claimed GRACE performance and allows the observation of, for example, gravity field changes due to continental hydrology (signal magnitude also indicated in Fig. 6, (Gruber et al. 2011)). In this preliminary analysis, we have not analyzed the possibilities to decouple the signature of the spacecraft's velocity and gravitational potential from the frequency comparison measurements. Such an approach, as is discussed by Dirkx et al. (2016) for planetary laser ranging data, could be used to quantify more robustly the influence of velocity uncertainties on potential measurements using clocks. Conversely, such an analysis would quantify the degree to which the relativistic Doppler shift can be used to dynamically determine the spacecraft's state when making use of clock comparisons.

8 Discussion and Conclusions

Various geodetic applications of highly accurate and stable clocks at the 10^{-18} level for the relative frequency shift (corresponding to 1 cm in height) have been addressed. We discussed both the prospects of implementing in practice the new clock-based measurement concepts and their challenges. Our firm conclusion is that the quantum clocks and time metrology will provide us with unique opportunities to support and largely extend the tools for gravity field measurements and physical height determination based on the direct access to the gravity potential differences.

We revisited the relevant time scales used in geodesy and showed the existence of growing inconsistencies in their definition and realization at a 10^{-18} accuracy level. This mainly concerns the use of the (non-constant) gravity potential value W_0 on the geoid, and its relation to the international time scales TT and TAI. In a foreseeable future, clarification and re-definition of the currently-used time concepts will be required to make them consistent with the highly-precise clock measurements.

We have provided a set of formulas to directly use the observed frequency shifts for the determination of physical height differences, where the relative clock measurements replace the differences of geopotential numbers. Applying this conceptually-new approach can facilitate or even fully resolve the problem of the systematic discrepancies existing between various height systems on the regional, national and intercontinental level. Clocks can connect tide gauges through satellite and/or fiber links providing the vertical datum of national height systems and help to determine the present offsets between them. Even discrepancies between independent geoid solutions stemmed from applying different classical techniques (GNSS/levelling versus gravimetric methods) can be determined and eliminated by making use of the precise chronometric measurements. These discrepancies are known to reaching centimeters to decimeters and can well be recovered already with the present-day clock technology. At the envisaged cm level of accuracy, time-variable parts affecting the clock measurements due to observer's height, position or Earth's density variations have to be considered. The most prominent effect is caused by the solid Earth tides but smaller variations, such as ocean tides or inaccurate knowledge of the Earth rotation parameters, play a significant role and are to be taken into consideration in the data processing algorithms.

The direct use of observed gravity potential values for regional gravity field recovery has been studied as a test case in the Massif Central region crossing a number of mountain ranges. We have clearly demonstrated in our simulation scenario that adding the clock-based potential values to the existing data set would notably improve the final gravity field solution. The bias to a reference solution and total rms error could be reduced remarkably well, up to a few orders of magnitude. Another important conclusion stemming from our simulations is that in solving the problem of gravity field recovery it is not required to have a dense clock network, only a very few percent of clock measurements compared to the number of needed gravity data is sufficient.

The possible use of ultra-precise space-borne clocks has been revisited for deriving potential height differences at satellite altitude and using these satellite data for recovering the global gravity field. It has been shown that extremely challenging requirements have to be met in order to use space-borne clocks to successfully fulfill this task. In order to be able to exploit the gravity field information content of clocks with a stability of 10^{-18}, precise reference clocks are needed along with two-way frequency transfer for eliminating the Doppler effect correction errors. In addition, the position and velocity of the clock need to be known with a very high precision. The current technological level of GNSS receivers allow for determining the position with the required cm-level precision. However, the velocity needs to be determined with such a high precision, that the associated velocity perturbations caused by the gravity field anomalies might be used as well as a fundamental observable for determining Earth's gravity field. Finally, the 10^{-18} clock stability needs to be achievable for very short integration times of the order of seconds. Even if this would be possible and achieved, the existing space geodesy concepts such as GRACE-type ll-SST lead to a better observability of Earth's time variable gravity. However, our study does show that the frequency transfer may be used to improve the determination of the spacecraft velocity.

Acknowledgements The International Space Science Institute (ISSI), is acknowledged for having provided the opportunity of presenting the work described in this paper at the Workshop on High Performance Clocks, 30 November - 4 December 2015 in Bern, Switzerland. Jürgen Müller gratefully acknowledges support by the DFG Sonderforschungsbereich (SFB 1128: geo-Q) Relativistic Geodesy and Gravimetry with Quantum Sensors. We are also thankful to two anonymous referees whose peer review and critical comments helped us to significantly improve the presentation of the manuscript.

References

D.C. Agnew, Earth tides, in *Treatise on Geophysics*, ed. by G. Schubert (Elsevier, Amsterdam, 2007), pp. 163–195. doi:10.1016/B978-044452748-6.00056-0

L. Blanchet, C. Salomon, P. Teyssandier, P. Wolf, Relativistic theory for time and frequency transfer to order c^{-3}. Astron. Astrophys. **370**, 320–329 (2001)

V.A. Brumberg, S.M. Kopejkin, *Relativistic Theory of Celestial Reference Frames* (Springer, Netherlands, Dordrecht, 1989a), pp. 115–141. doi:10.1007/978-94-009-0933-5_6

V.A. Brumberg, S.M. Kopejkin, Relativistic reference systems and motion of test bodies in the vicinity of the earth. Il Nuovo Cimento B **103**(1), 63–98 (1989b). doi:10.1007/BF02888894

M. Burša, S. Kenyon, J. Kouba, Z. Šíma, V. Vatrt, V. Vítek, M. Vojtíšková, The geopotential value W_0 for specifying the relativistic atomic time scale and a global vertical reference system. J. Geod. **81**(2), 103–110 (2007)

D. Calonico, A. Cina, F. Levi, I.H. Bendea, L. Lorini, A. Godone, Gravitational redshift at INRIM. Metrologia **44**(5), 44–48 (2007)

O. Carraz, C. Siemes, L. Massotti, R. Haagmans, P. Silvestrin, A spaceborne gradiometer concept based on cold atom interferometers for measuring earth's gravity field. Microgravity Sci. Technol. **26**, 139–145 (2015). doi:10.1007/s12217-014-9385-x

N. Chiodo, K. Djerroud, O. Acef, A. Clairon, P. Wolf, Lasers for coherent optical satellite links with large dynamics. Appl. Opt. **52**, 7342–7351 (2013)

S.J. Claessens, C. Hirt, Ellipsoidal topographic potential: new solutions for spectral forward gravity modeling of topography with respect to a reference ellipsoid. J. Geophys. Res., Solid Earth **118**(11), 5991–6002 (2013). doi:10.1002/2013JB010457

O.L. Colombo, *The Global Mapping of Gravity with Two Satellites* (1984)

C. Dahle, F. Flechtner, C. Gruber, D. König, R. König, G. Michalak, K.-H. Neumayer, GFZ GRACE level-2 processing standards document for level-2 product release 0005. Scientific technical report STR12/02—data, revised edn., Helmholtz-Zentrum Potsdam, Deutsches Geoforschungszentrum, 2013

T. Damour, M. Soffel, C.-M. Xu, General-relativistic celestial mechanics. i. method and definition of reference systems. Phys. Rev. D **43**, 3273–3307 (1991). doi:10.1103/PhysRevD.43.3273

T. Damour, M. Soffel, C.-M. Xu, General-relativistic celestial mechanics, II: translational equations of motion. Phys. Rev. D **45**, 1017–1044 (1992). doi:10.1103/PhysRevD.45.1017

T. Damour, M. Soffel, C.-M. Xu, General-relativistic celestial mechanics, III: rotational equations of motion. Phys. Rev. D **47**, 3124–3135 (1993). doi:10.1103/PhysRevD.47.3124

N. Dayoub, S.J. Edwards, P. Moore, The Gauss-Listing geopotential value W_0 and its rate from altimetric mean sea level and GRACE. J. Geod. **86**(9), 681–694 (2012)

H. Denker, Regional gravity field modeling: theory and practical results, in *Sciences of Geodesy-II*, ed. by G. Xu (2013), pp. 185–291. doi:10.1007/978-3-642-28000-9_5

H. Denker, Height differences observed by different geodetic methods. Private communication, 2016

H. Denker, L. Timmen, C. Voigt, Gravity field modelling for optical clock comparisons, 2015

D. Dirkx, R. Noomen, P.N.A.M. Visser, L. Gurvits, L.L.A. Vermeersen, Space-time dynamics estimation from space mission tracking data. Astron. Astrophys. **587 (A156)**, 1–10 (2016). doi:10.1051/0004-6361/201527524

K. Djerroud, O. Acef, A. Clairon, P. Lemonde, C.N. Man, E. Samain, P. Wolf, Coherent optical link through the turbulent atmosphere. Opt. Lett. **35**, 1479–1481 (2010)

M. Drinkwater, R. Haagmans, D. Muzzi, A. Popescu, R. Floberghagen, M. Kern, M. Fehringer, The GOCE gravity mission: ESA's first core explorer, in *3rd GOCE User Workshop*, Frascati, Italy, 6–8 November 2006 (2007), pp. 1–7

C. Dunn, W. Bertiger, Y. Bar-Sever, S. Desai, B. Haines, D. Kuang, G. Franklin, I. Harris, G. Kruizinga, T. Meehan, S. Nandi, D. Nguyen, T. Rogstad, J.B. Thomas, J. Tien, L. Romans, M. Watkins, S.-C. Wu, S. Bettadpur, J. Kim, Instrument of Grace: GPS augments gravity measurements, in *GPS World* (2003), pp. 16–28

ESA, Earth observation science strategy for ESA—a new era for scientific advances and societal benefits. ESA SP-1329/1&2, 2015

V.F. Fateev, S.M. Kopeikin, S.L. Pasynok, Effect of irregularities in the earth's rotation on relativistic shifts in frequency and time of earthbound atomic clocks. Meas. Tech. **58**(6), 647–654 (2015). doi:10.1007/s11018-015-0769-0

V.A. Fock, *The Theory of Space, Time and Gravitation*, 2nd revised edn. (Macmillan, New York, 1964), translated from the Russian by N. Kemmer. 1st edn.: Pergamon, New York, 1959

R. Forsberg, A new covariance model for inertial gravimetry and gradiometry. J. Geophys. Res. **92**, 1305–1310 (1987). doi:10.1029/JB092iB02p01305

R. Forsberg, C.C. Tscherning, *An Overview Manual for the GRAVSOFT* (Univ. of Copenhagen Press, Copenhagen, 2008)

C. Förste, S. Bruinsma, O. Abrikosov, F. Flechtner, J.-C. Marty, J.-M. Lemoine, C. Dahle, H. Neumayer, F. Barthelmes, R. König, R. Biancale, EIGEN-6C4—the latest combined global gravity field model including GOCE data up to degree and order 1949 of GFZ Potsdam and GRGS Toulouse, in *EGU General Assembly Conference Abstracts*, vol. 16 (2014), p. 3707

L.-L. Fu, Determining ocean circulation and sea level from satellite altimetry: progress and challenges, in *Oceans from Space*, ed. by V. Barale, J.F.R. Gower, L. Alberotanza (Springer, New York, 2010), pp. 147–163. doi:10.1007/978-90-481-8681-5_9

J. Gersl, P. Delva, P. Wolf, Relativistic corrections for time and frequency transfer in optical fibres. Metrologia **52**, 552–564 (2015). doi:10.1088/0026-1394/52/4/552

F.R. Giorgetta, W.C. Swann, L.C. Sinclair, E. Baumann, I. Coddington, N.R. Newbury, Optical two-way time and frequency transfer over free space. Nat. Photonics **7**, 434–438 (2013)

H. Goldstein, C. Poole, J. Safko, *Classical Mechanics* (Addison-Wesley, San Francisco, 2002)

Th. Gruber, J. Bamber, M.F.P. Bierkens, H. Dobslaw, M. Murböck, M. Thomas, L.P.H. van Beek, T. van Dam, L.L.A. Vermeersen, P.N.A.M. Visser, Simulation of the time-variable gravity field by means of coupled geophysical models. Earth Syst. Sci. Data **3**, 19–35 (2011). doi:10.5194/essd-3-19-2011

W.A. Heiskanen, H. Moritz, *Physical Geodesy* (Freeman, San Francisco, 1967)

J. Ihde, R. Barzaghi, U. Marti, L. Sánchez, M. Sideris, H. Drewes, C. Foerste, T. Gruber, G. Liebsch, R. Pail, Report of the ad hoc group on an international height reference system (IHRS). IAG Reports 2011–2015, Travaux de l'AIG, vol. 39, 2015. http://iag.dgfi.tum.de/index.php?id=329

Jet Propulsion Laboratory, GRACE follow-on. http://gracefo.jpl.nasa.gov/mission/mission3, 2017
S.M. Kopejkin, Celestial coordinate reference systems in curved space-time. Celest. Mech. **44**, 87–115 (1988). doi:10.1007/BF01230709
S.M. Kopejkin, Relativistic manifestations of gravitational fields in gravimetry and geodesy. Manuscr. Geod. **16**(5), 301–312 (1991)
S. Kopeikin, M. Efroimsky, G. Kaplan, *Relativistic Celestial Mechanics of the Solar System* (Wiley, Berlin, 2011)
S. Kopeikin, W. Han, E. Mazurova, Post-Newtonian reference ellipsoid for relativistic geodesy. Phys. Rev. D **93**(4), 044069 (2016). doi:10.1103/PhysRevD.93.044069
S.P. Kuzin, S.K. Tatevian, S.G. Valeev, V.A. Fashutdinova, Studies of the geocenter motion using 16-years DORIS data. Adv. Space Res. **46**, 1292–1298 (2010). doi:10.1016/j.asr.2010.06.038
G. Lion, I. Panet, P. Wolf, C. Guerlin, S. Bize, P. Delva, Determination of a high spatial resolution geopotential model using atomic clock comparisons. J. Geod. (2017). doi:10.1007/s00190-016-0986-6
E. Mai, J. Müller, General remarks on the potential use of atomic clocks in relativistic geodesy. Z. Geoinformation Landmanagement **4**, 257–266 (2013)
R. Mayrhofer, R. Pail, Future satellite gravity field missions: feasibility study of post-Newtonian method, in *Geodesy for Planet Earth*, ed. by S. Kenyon et al. International Association of Geodesy Symposia, vol. 136 (Springer, Switzerland, 2012), pp. 231–238. doi:10.1007/978-3-642-20338-1_28
H. Moritz, *Advanced Physical Geodesy* (Herbert Wichmann Verlag, Karlsrue, 1980)
J. Müller, Erdmessung mit Quanten und Relativität, in *BWG Jahrbuch* (2016). arXiv:1608.08407
R. Pail, H. Goiginger, W.-D. Schuh et al., Combined satellite gravity field model GOCO01S derived from GOCE and GRACE. Geophys. Res. Lett. **37**(20), L20314 (2010). doi:10.1029/2010GL044906
N.K. Pavlis, M.A. Weiss, The relativistic redshift with 3×10^{-17} uncertainty at NIST, Boulder, Colorado, USA. Metrologia **40**(2), 66–73 (2003)
G. Petit, B. Luzum, IERS conventions (2010). IERS technical note 36, 2010
G. Petit, P. Wolf, Computation of the relativistic rate shift of a frequency standard. IEEE Trans. Instrum. Meas. **46**(2), 201–204 (1997)
G. Petit, P. Wolf, P. Delva, Atomic time, clocks and clock comparisons in relativistic space-time: a review, in *Frontiers in Relativistic Celestial Mechanics, vol. 2: Applications and Experiments* (de Gruyter, Berlin, 2014), pp. 249–283
P. Rebischung, Z. Altamimi, T. Springer, Insensitivity of GNSS to geocenter motion through the network shift approach (invited), in *AGU Fall Meeting Abstracts* (2013)
A.S. Richey, B.F. Thomas, M.-H. Lo, J.T. Reager, J.S. Famiglietti, K. Voss, S. Swenson, M. Rodell, Quantifying renewable groundwater stress with GRACE. Water Resour. Res. **51**(7), 5217–5238 (2015). doi:10.1002/2015WR017349
L. Sánchez, Towards a vertical datum standardisation under the umbrella of global geodetic observing system. J. Geod. Sci. **2**, 325–342 (2012). doi:10.2478/v10156-012-0002-x
L. Sánchez, Ein einheitliches vertikales Referenzsystem für Südamerika im Rahmen eines globalen Höhensystems. PhD thesis, Technische Universität Dresden, 2015. http://nbn-resolving.de/urn:nbn:de:bsz:14-qucosa-162350
E.J.O. Schrama, Gravity field error analysis: applications of gps receivers and gradiometers on low orbiting platforms. J. Geophys. Res. **96**(B12), 20,041–20,051 (1991)
B. Sheard, G. Heinzel, K. Danzmann, D. Shaddock, W. Klipstein, W. Folkner, Intersatellite laser ranging instrument for the GRACE follow-on mission. J. Geod. **86**(12), 1083–1095 (2012)
B. Simon, A. Lemaitre, J. Souchay, Oceanic tides, in *Tides in Astronomy and Astrophysics*, ed. by J. Souchay, S. Mathis, T. Tokieda (Springer, Berlin, 2013), pp. 83–114. doi:10.1007/978-3-642-32961-6_3
D.A. Smith, There is no such thing as "the" egm96 geoid: subtle points on the use of a global geopotential model. I.Ge.S. Bull. **8**, 17–28 (1998)
W.H.F. Smith, D.T. Sandwell, Global sea floor topography from satellite altimetry and ship depth soundings. Science **277**(5334), 1956–1962 (1997). doi:10.1126/science.277.5334.1956
M. Soffel, S.A. Klioner, G. Petit, P. Wolf, S.M. Kopeikin, P. Bretagnon, V.A. Brumberg, N. Capitaine, T. Damour, T. Fukushima, B. Guinot, T.-Y. Huang, L. Lindegren, C. Ma, K. Nordtvedt, J.C. Ries, P.K. Seidelmann, D. Vokrouhlicky, C.M. Will, C. Xu, The IAU 2000 resolutions for astrometry, celestial mechanics, and metrology in the relativistic framework: explanatory supplement. Astron. J. **126**, 2687–2706 (2003)
B.D. Tapley, S. Bettadpur, J.C. Ries, P.F. Thompson, M.M. Watkins, GRACE measurements of mass variability in the earth system. Science **305**, 1503–1505 (2004)
Thales Alenia Spazio, DEOS, IAPG, GIS, ULUX, DEIMOS, and ONERA, Assessment of a next generation mission for monitoring the variations of earth's gravity. SD-RP-AI-0688, final report, ESTEC contract No. 22643/09/NL/AF, European Space Agency, December 2010
W. Torge, J. Müller, *Geodesy*, 4th edn. (De Gruyter, Berlin/Boston, 2012)

I. Velicogna, J. Wahr, Acceleration of Greenland ice mass loss in spring 2004. Nature **443**, 329–331 (2006a). doi:10.1038/nature05168

I. Velicogna, J. Wahr, Measurements of time-variable gravity show mass loss in Antarctica. Science **311**, 754–1756 (2006b). doi:10.1126/science.112378

P.N.A.M. Visser, J. van den IJssel, R. Koop, R. Klees, Exploring gravity field determination from orbit perturbations of the European gravity mission GOCE. J. Geod. **75**(2/3), 89–98 (2001)

P.N.A.M. Visser, N. Sneeuw, C. Gerlach, Energy integral method for gravity field determination from satellite orbit coordinates. J. Geod. **77**(3/4), 207–216 (2003)

C. Voigt, H. Denker, L. Timmen, Investigation of time-variable components of the gravity potential for optical clock comparisons, in *26th General Assembly of International Union of Geodesy and Geophysics (IUGG)*, Prague, Czech Republic (2015)

C. Voigt, H. Denker, L. Timmen, Time-variable gravity potential components for optical clock comparisons and the definition of international time scales. Metrologia **53**(6), 1365–1383 (2016)

P. Wolf, G. Petit, Relativistic theory for clock syntonization and the realization of geocentric coordinate times. Astron. Astrophys. **304**, 654–661 (1995)

X. Wu, J. Ray, T. van Dam, Geocenter motion and its geodetic and geophysical implications. J. Geodyn. **58**, 44–61 (2012). doi:10.1016/j.jog.2012.01.007

Space Sci Rev (2018) 214:22
https://doi.org/10.1007/s11214-017-0457-2

The Application of Coherent Local Time for Optical Time Transfer and the Quantification of Systematic Errors in Satellite Laser Ranging

K. Ulrich Schreiber[1] · **Jan Kodet**[1]

Received: 19 November 2016 / Accepted: 7 December 2017 / Published online: 20 December 2017

Abstract Highly precise time and stable reference frequencies are fundamental requirements for space geodesy. Satellite laser ranging (SLR) is one of these techniques, which differs from all other applications like Very Long Baseline Interferometry (VLBI), Global Navigation Satellite Systems (GNSS) and finally Doppler Orbitography and Radiopositioning Integrated by Satellite (DORIS) by the fact that it is an optical two-way measurement technique. That means that there is no need for a clock synchronization process between both ends of the distance covered by the measurement technique. Under the assumption of isotropy for the speed of light, SLR establishes the only practical realization of the *Einstein Synchronization* process so far. Therefore it is a powerful time transfer technique. However, in order to transfer time between two remote clocks, it is also necessary to tightly control all possible signal delays in the ranging process. This paper discusses the role of time and frequency in SLR as well as the error sources before it address the transfer of time between ground and space. The need of an improved signal delay control led to a major redesign of the local time and frequency distribution at the Geodetic Observatory Wettzell. Closure measurements can now be used to identify and remove systematic errors in SLR measurements.

Keywords Satellite laser ranging · Optical time transfer · Accurate time and frequency

High Performance Clocks with Special Emphasis on Geodesy and Geophysics and Applications to Other Bodies of the Solar System
Edited by Rafael Rodrigo, Véronique Dehant, Leonid Gurvits, Michael Kramer, Ryan Park, Peter Wolf and John Zarnecki

✉ K.U. Schreiber
ulrich.schreiber@bv.tum.de

[1] FESG, Geodätisches Observatorium Wettzell, Technical University of Munich, Sackenrieder Str. 25, 93444 Bad Kötzting, Germany

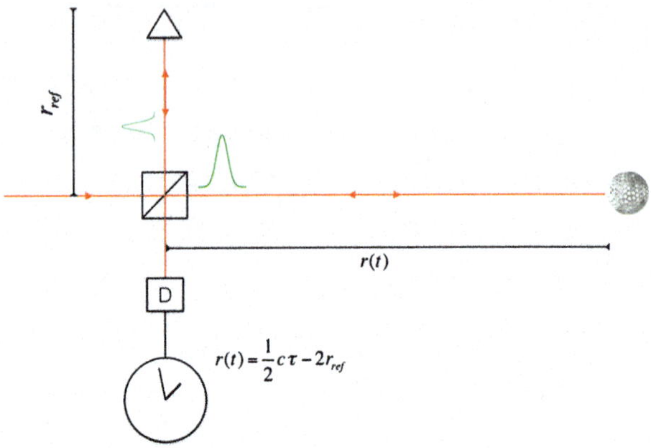

Fig. 1 Simplified illustration of the satellite laser ranging technique. All signal delays, which are not part of either the measured distance or the calibration range, cancel out entirely

1 The Technique of Satellite Laser Ranging

1.1 A Two-Way Concept of Measuring the Time of Flight

Satellite Laser Ranging (SLR) establishes the distance between a fixed point (station) on the ground and a satellite target, utilizing a time of flight measurement of short laser pulses. It is unique in the ensemble of space geodetic measurement techniques, because it operates in the optical regime and it realizes a two-way measurement concept. This means that there is no need to synchronize the transmitter and receiver clocks in a similar fashion as necessary for the VLBI and the GNSS techniques, because the time of flight of short mode-locked laser pulses is established by the same local clock, so that a clock offset cancels out. Although the individual instrumental components of SLR require challenging technologies, the underlying measurement concept is very simple and also has a clear unambiguous geometrical meaning (Degnan 1993). Figure 1, drawn in the form reminiscent of a Michelson interferometer, illustrates this process. Short laser pulses are generated by a mode-locked laser and a small portion of this signal is directed to the start detector, timing the epoch t_1 of the outgoing signal. Technically this requires a beam splitter and in the interest of simplicity, we have placed this device at the reference point of our measurement system. Most of the laser pulse intensity is steered towards the satellite of interest by the help of an optical telescope not shown in the functional diagram. The laser pulses are back reflected from a suitable retro reflector array and also timed as t_2 on the local clock of the SLR station. With $\tau = t_2 - t_1$ the distance to the satellite is obtained as $r(t) = \frac{1}{2}c\tau - 2r_{ref}$, where r_{ref} is the contribution of the reference arm, providing a range of precisely known length. We note that for each laser shot this detection process happens at an delayed instance, which is not a priori known. It is caused by the travel time of the laser pulses and the electrical signals inside the system. Hence it is necessary to compensate these extra delays by performing a reference measurement over an exactly known distance. In our simplified diagram this corresponds to the reference arm of the sketched interferometer. From the idealized situation shown in Fig. 1 one can see that every delay cancels out. In the real instrumentation a few compromises are necessary for technical reasons, so that the ideal situation is only approximated.

SLR systems typically operate solid state lasers, where the active material is Nd:YAG (an yttrium aluminium garnet crystal doped with neodymium), which produce short laser pulses with a pulse width of the order of 10 ps $\leq \tau \leq$ 200 ps at a wavelength of 1.064 μm. Optical frequency doubling places the wavelength at 532 nm, which is in the middle of the visible part of the optical spectrum, where photodetectors are most sensitive with quantum efficiencies exceeding 70%. When the laser pulse is generated it is time tagged as t_0 on the local clock and that determines the epoch of the measurement. The laser pulse is then guided to a telescope, which can be steered such that it points to a geodetic satellite and tracks it along its orbit, while the satellite is above the horizon of the observing station. The intensity of the satellite reflection is highly variable and more than 10 orders of magnitude smaller when it hits the detector. Therefore a high detector gain is required to produce a measurable signal response at t_1. For practical reasons all currently operating SLR systems separate the process of the range measurement from the process of the reference measurement. Sometimes the calibration of the SLR system is only repeated every hour or even less often. The time interval Δt of the measurement between the epoch of the return signal and the epoch of the laser fire event contains a number of system and concept dependent delays, which have to be removed before the exact distance between the reference point on the ground (namely the intersection point of the horizontal and the vertical axis of the telescope) and the reflector array on the satellite can be derived.

$$\Delta t_e^s\big|_{t_0} = \frac{2}{c}\left(\rho_e^s + \delta\rho_{atm} + \delta\rho_{rel}\right) + \frac{1}{c}\delta\rho_{sys} + \epsilon_e^s \tag{1}$$

Here $\Delta t_e^s|_{t_0}$ is the time interval between laser fire and detection at the epoch t_0, ρ_e^s is the distance between the reference points, $\delta\rho_{atm}$ the atmospheric contribution to the path delay, $\delta\rho_{rel}$ accounts for relativistic effects and $\delta\rho_{sys}$ is the system delay. The value c is the vacuum velocity of light and ϵ_e^s corresponds to stochastic measurement errors. In order to relate the optical distance measurement to the center of mass of the satellite, the eccentricity vector between the center of mass of the satellite and the reflector array has to be applied. While this correction is fairly simple for spherically shaped satellites with laser reflectors all over the surface like for the satellites LAGEOS, STARLETTE or ETALON, this correction can be very tedious and attitude dependent for irregular shaped Earth observing satellites with an reflector array somewhere at a corner. This problem can be even worse when the satellite is losing mass throughout its mission for example by burning fuel.

1.2 The Importance of Time and Frequency in SLR

For SLR measurements we have to distinguish between the time used in the context of establishing the epoch of an individual range measurement and the actual round trip time interval measurement of the laser pulse. The former is related to UTC time, while the latter is obtained from the frequency of the oscillator of the clock. This measurement epoch is associated with the position of the satellite along the orbit and the requirements for the accuracy of this value are not particularly high. A low Earth orbiting satellite with velocities of the order of 10 km/s, changes its position approximately by 1 cm in 1 µs. So, if the station clock is offset from true UTC by several µs, the satellite appears shifted along its orbit. Radiation pressure and more importantly a deceleration from the gas pressure of the residual atmosphere are also affecting the satellite motion and have to be modelled in the analysis process. Therefore the correct measurement epoch (station clock offset) is finally established from the non-linear least squares adjustment procedure from the combined measurements of a network of SLR stations.

The accurate measurement of the round trip time of the light is far more critical for the measurement process than the epoch, because the error margin of the established distance depends on the resolution of the timing process of measuring the time of flight of a laser

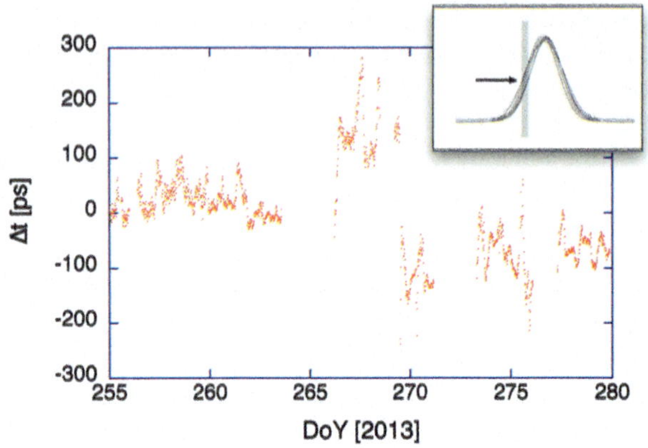

Fig. 2 The variation of the time delay between the rising edges of the pulse per second signals, derived from the same maser in 2 different buildings. The *inset* illustrates schematically the detection timing jitter in the presence of unavoidable phase noise. The *arrow* indicates the detection threshold and the *greyed out area* illustrates the uncertainty in timing

pulses. This in turn is depending on the pulse width of the laser, the bandwidth of the detector and signal conditioning unit and the granularity of the timescale of the clock used in the timing process. Modern detectors provide signal rise times of about 100 ps, a typical laser generates pulses in the range of 30 ps FWHM and the smallest division of modern event timing devices is just below of 3 ps. As the measurement precision improves with averaging, it is therefore desirable to operate SLR systems with a high repetition rate of 1–2 kHz in order to exploit averaging from a high number of range measurements. However, the potential gain from going to higher repetition rates is not indefinite, because the range measurements do not follow a Gaussian distribution as outlined in the next section. As a consequence of such highly developed subsystems we find a measurement precision of the order of 3 mm for a modern SLR systems.

2 Error Sources and Limitations in SLR

2.1 Temperature Related Effects on the Timing Electronics

While a single shot precision of less than 15 ps is routinely achieved by several SLR systems, the respective obtained accuracy for the reconstruction of the satellite orbit is not better than 20–45 ps. This is not only because of some fundamental short-comings of the current realization of the SLR technique (Appleby et al. 2016). Apart from small deficiencies in the modelling of the satellite dynamics and the contribution of the chosen frame of reference a number of small technically induced systematic biases need to be addressed, so that this complex error budget eventually can be disentangled. Together with stability and bandwidth related issues of comparable size within the VLBI technique, this leads to an apparent discrepancy for the scale of the ITRF of the order of 2 parts in 10^9. Several mechanisms are contributing to this error budget. Time corresponds to the integral of a reference frequency of a stable clock. Timing signals therefore are typically derived from the leading edge of electrical pulses. In order to obtain a high measurement resolution this ideally requires an infinite steep slope, which is practically not achievable. The maximum bandwidth obtainable from the available technology of electronic circuitry is between 1 and 10 GHz. Therefore the actual timing accuracy for the threshold detection is highly dependent on the stability of the ground potential and the bandwidth of the measurement channel, as well as the phase noise on the rising edge of the signal. Figure 2 illustrates this for a comparison of the pulse

Fig. 3 The variation of the measured time interval along a short fixed local calibration range for a constant optical signal level. The variability of the range is due to the unavoidable and variable signal propagation delay inside the electronic units of the ranging system

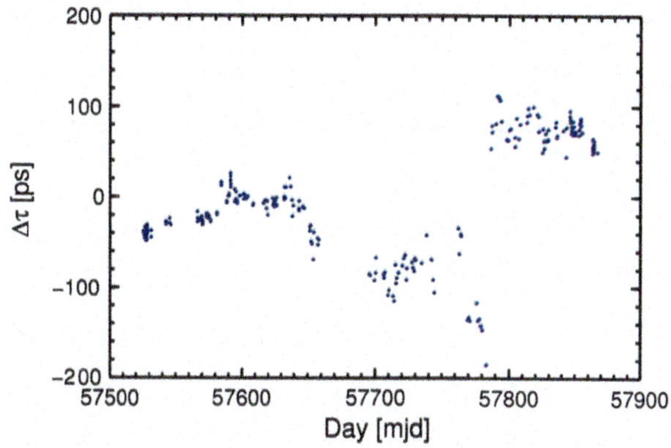

per second (PPS) square wave signals between two buildings approximately 60 m apart. The inset in Fig. 2 illustrates how fluctuations in ground potential and phase noise map into time in the presence of a bandwidth limited detection. However, the most dramatic impact on the displayed measurement time series is caused by the temperature dependence of the transition delay in the applied electronic circuitry. Evidence to this is the diurnal signature of measured time delays in the first part of Fig. 2. Technical problems of the air conditioning system and human activity around the clock area led to the much larger excursions for the delay of the clock comparison, shown in the second part of the plot, before the delay returns to the original pattern towards the end of this 25 day observation period.

While Fig. 2 illustrates the difficulty to obtain a long term stable delay comparison between two buildings, Fig. 3 depicts the achievable long term system stability of the calibration of a single SLR station. Each data point in the diagram corresponds to the mean value of a measurement series, where the laser is fired a few thousand times at a flat calibration target. For each shot the epoch of the laser fire event is taken by the start detector. Shortly afterwards the calibration stop is obtained from the telescope detector, when the laser pulse is returned to the system after passing through the reference range. Since the reference range and the detectors are held constant, Fig. 3 provides the stability of the timing circuitry. Over a period of one year, one can see a peak to peak variation of about 300 ps in the electronic circuitry of the system delay. The main contributor for this variability are temperature induced delays in the electronics. For most of the time the change in value from one day to the next is small, however there are also some sudden jumps in the data. Since the start and stop signals are timed on the same clock, a clock offset does not affect the measurements, because it is common to both timing processes and disappears in the difference. Keeping in mind that the entire measurement error should not exceed a few ps, this also means that not only the clock must be the same, but also the electronic signal path for both, laser fire (start signal) and echo detection (stop signal). A different signal path for timing the start and the stop event inside the timer hardware of an SLR system introduces an unknown and due to the discussed temperature sensitivity slowly variable delay, which consequently compromises the accuracy of the measurement.

2.2 Delays from the Conversion from the Optical to the Electrical Domain

So far we have discussed variable system delays, which are mainly caused by temperature related variations in the transfer of broadband timing signals along the electronic circuitry of the event timing system. While this type of system delay is caused on the signal path

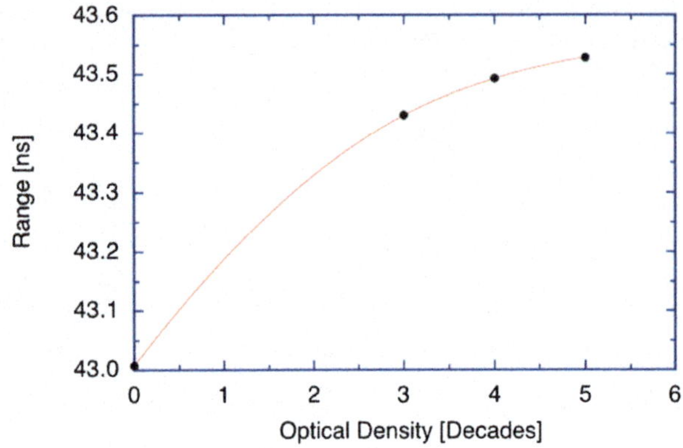

Fig. 4 This figure illustrates the dependence of the conversion delay of an optical input signal of a photodiode to an electrical output signal on the input signal intensity. Over 5 orders of magnitude of variation of the input intensity caused a non-linear change in detector timing delay of more than 500 ps

between the various components of the ranging system, the optical echo detection process also contributes to the error budget. The response of a bound electron to the electromagnetic force of light is fast but not instantaneous (Hassan et al. 2016). While the resultant delay is on the sub-fs timescale in the interaction of electrons to light, the macroscopic response, namely the time delay between the arrival of a short light pulse and the appearance of the electrical output voltage on a fast linear detector is already of the order of 211 ps and for a high bandwidth photon counting device this builds up to about 1.745 ns (Prochazka et al. 2011), leading to a large gap with significant uncertainty between the moment of arrival of an optical signal at the detector and the actual electrical timing output response. The main dilemma is that the actual range measurements are done in the optical domain, while the detection and timing is done in the electrical domain. It is usually assumed that the conversion delay remains constant between the optical range measurement and the system calibration and therefore cancels out as a common mode effect. While this is a reasonable expectation for a controlled laboratory environment, this does not fully apply for satellite ranging, because the intensity of the return signal varies a lot from one satellite target to another and even during a single measurement sequence, as the transmission of the atmosphere can change quickly. Figure 4 illustrates this intensity dependent detector delay from an laboratory measurement. While the diagram in Fig. 4 also covers extreme intensity differences, a variation of the signal intensity of around 1 order of magnitude in the regime of low intensities (the right side of Fig. 4) are typical in satellite laser ranging. So uncompensated variable systematic errors of 10 ps and more may easily arise from intensity variations of the satellite return signal. Finally it also matters, where on the active area of the photodiode the avalanche process is seeded (Blazej and Prochazka 2009). The signal growth is faster in the center of the active area, where no restrictions are present. At the perimeter of the photosensitive area the signal build up time is longer by about 5–10 ps. Together with additional range errors caused by temporal return pulse spreading due to the geometric depth of the reflector arrays on the satellite targets, these effects currently limit the achievable accuracy and stability of the SLR technique, because these error sources are not present in the system calibration process. We note that the obtained orbit accuracy of the LAGEOS satellites currently remains at a level of about 45 ps (Appleby et al. 2016), although the precision of repeated satellite range measurements may be as low as 10 ps. A clear indication for the non-negligible effect of the discussed systematic errors is the fact that the satellite range measurements are not normally distributed, while the histograms of the calibration measurements exhibit a clean Gaussian shape.

2.3 Deficiencies in the Estimate of the Geometrical Reference

Another critical process is the determination of the geometrical reference point of an SLR system. While the ranging process as such is very elegant; we have a well defined signal source and a geometrically clearly defined endpoint (the retro reflector) and the measurement process itself follows Fermat's law of the shortest path (time), it is nevertheless very challenging to cope with technically necessary compromises. These arise most prominently from limitations of the dynamic range of the measurement signals. In extreme situations one finds a difference of the order of -170 dB between the generated and the detected laser pulse. As a consequence, transmit and receive functions have to be separated at the telescope. One way to overcome this difficulty is to split the signal path and use a transmit telescope different from the receive telescope or to introduce a switching unit of some making that connects the laser and the detector to the telescope in an alternating fashion. Either way, this requirement introduces a small asymmetry into the measurement process and it is not the only one. In order to establish the moment where the laser pulse passes the reference point of the SLR system on its way out to the satellite target, one has to measure the time that the laser pulse takes to travel from the laser to the telescope reference point, including the effect of cable delays and transfer times through electronic circuits on the timing of the pulse. To achieve this, SLR systems usually utilize a calibration process, which consists of repeated range measurements to a nearby geometrically flat ground target in a fixed and known position, assuming that the local signal delays are the same for both the satellite and the calibration measurement. This is almost true, but not perfectly. Since the laser signal is almost instantly back at the detector ($\Delta\tau \approx 100$ ns), there is a small asymmetry in the ranging system electronics to allow for this. Typical satellite range measurements have a time interval of several milliseconds between laser fire and echo arrival. On top of these small differences in the signal delays between calibration and range measurement one has also to deal with similar small problems caused by the geometrical depth of the satellite target, when there is more than a single cube corner reflector involved. Variations of the signal strength of the detected laser returns are yet another complication.

2.4 Timing Errors Affecting the Scale

The clock frequency, when it is not representing the SI second affects the range measurement in the form of an error in scale. When the generated timescale significantly deviates from the length of the SI second and the stability of its representation is not given, it can be a limitation for the accuracy of the measurements. However, a typical commercial GPS disciplined oscillator reduces this error already to less than 1 mm for distances corresponding to the LAGEOS orbit. So from a clock point of view scale errors are not of concern in SLR today. Table 1 summarizes all the discussed systematic error contributions in a compact form. It is important to note that the numbers given are only rough indications, because the SLR stations differ significantly from each other with respect to design and parameter setting. For example the pulse-width of one system may be 10 ps, while for another it is more like 200 ps. Some systems use semiconductor photo-detectors with fixed threshold detection, while others use photomultipliers with constant-fraction discrimination. All these small error sources, listed in Table 1, will result in a combined measurement bias, which is neither constant over time, nor will it be the same for each station, as many parameters such as laser pulse length, electrical ground potential, instantaneous laser power and detection bandwidth vary from one observing station to the next and from one day to the other. Together with small modelling errors, reaching from small orbit model deficiencies to an

Table 1 This table summarizes where potential systematic error sources exist in SLR systems. Since the design of the various SLR stations in the ILRS differ significantly from each other, the numerical values will also differ from one system to another

Error source	Range	Cause
Target depth	$\approx$ 3 mm (Lageos)	Difference between leading edge and centroid detection
	25 mm (Apollo 15)	
Atmosphere	$\approx$ 3 mm	Error in the determination of the air pressure on the ground (elevation dependence)
Photo-detection	0–12 mm	– Intensity variations (highly variable)
		– Non-reciprocal setup between calibration and ranging (slowly changing)
Pulse shape	$\approx$ 2 mm	Non-Gaussian signal distribution
Trigger Jitter	$\pm$2 mm	– Trigger threshold jitter (lead. edge detection)
		– Differential part of ground potential fluctuation between ranging and calibration (variable with time)
		– Near field versus far field laser pattern
Reference Point	0–30 mm (?)	Error in the determination of the system reference point (constant)
Clock	–	No concern

unfavorable distribution of data points along the orbit for the non-linear least squares adjustment process, these measurement delays add up such, that for example today's LAGEOS orbit is established with an uncertainty of the order of 2 cm by the global network, although the respective precision of the averaged measurements of many SLR stations are below 5 mm. Provided that the coherence of time can be increased such that it encompasses an entire geodetic observatory, the stable phase of the clock can be used to perform closure measurements within and across the techniques of space geodesy. This will then allow to perform accurate optical time transfer and as an additional advantage it also provides a powerful tool to establish and eradicate the otherwise not detectable systematic (delay) errors within each of the measurement techniques.

3 Local Two-Way Time and Frequency Distribution

3.1 The Current Limitations for the Distribution of Time and Frequency

Coherent time transfer from one clock to another requires a perfect control over the phase of the clock oscillator along the complete path between the two clocks who are compared. In the previous section we have discussed the technical difficulties that arise when clocks are to be compared at the level of a few ps on a single time of flight measurement. In this section we describe a way to mitigate the effect of these delays by a closed loop two-way delay compensation. The Geodetic Observatory Wettzell operates several hydrogen masers and an ensemble of 5 Cs clocks and contributes to the UTC timescale. It provides both time and frequency to all the space geodetic instrumentation of the observatory, namely 3 VLBI antennas, 2 SLR systems, several GNSS receivers and operates a dedicated time laboratory. The space geodetic instrumentation of the observatory is distributed over an area of several hectare, with 50–200 m distance between them. This means that signal lines for time and frequency have to be routed to the individual measurement devices, which introduces extra

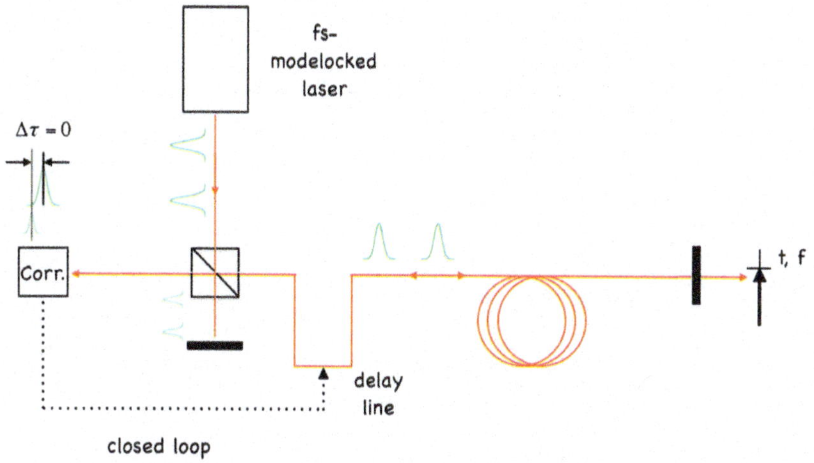

Fig. 5 Illustration of the lossless two-way time distribution system drawn in the way of an unequal arm Michelson interferometer. At the correlation stage the pulses returning from passing through the entire fiber link are superimposed on the locally reflected pulses. An actively controlled delay line is used to ensure the overlap of the local and remote pulses at all times

time delays and signal degradation artefacts, such as a reduction in signal bandwidth due to long cables for example. It also implies that we have to deal with variable extra delays in order to eventually relate the space clock to a ground clock with high accuracy. We recall that time corresponds to the integrated frequency and that the transfer of time requires a high bandwidth in order to locate the instantaneous phase with high resolution. Long cables between buildings act like a low pass filter on fast rising leading edge voltages of time markers. In the presence of different ground potentials and temperature variations in cables and even more so in the applied electronic devices through which the time signals are passing, highly variable transit delays develop, which are difficult to control. The delay between the various systems in Wettzell typically varies by as much as 1 ns over periods of several days with the offset in time mostly undetermined. Accurate time transfer, referenced to a distant master clock, requires a much stricter approach.

3.2 An All Optical Two-Way Time and Frequency Distribution

We have designed a local two-way time and frequency distribution, based on a femtosecond pulse laser, that can provide accurate time and stable reference frequencies to all measurement systems. Figure 5 illustrates the basic concept for a single link.

The reference frequency of 100 MHz from a stable hydrogen maser is locked to the repetition rate of the mode-locked pulse laser, generating equidistant ultrashort laser pulses of less than 100 fs pulse duration. These pulses are used to mark the zero phase instant on each cycle of the frequency. One of these pulses is associated with the beginning of the second in this time scale and this pulse is marked with an optical signal on a different wavelength as the pulse per second signal (PPS) in order to trigger all subsequent timing devices to the same timescale.

The pulse train is fed into an unequal arm Michelson type interferometer. The long arm is linking the time laboratory to the SLR facility via a fiber. A combination of the more than 100 m long fiber link and a coil of a conjugate fiber of the same length ensure that the pulse

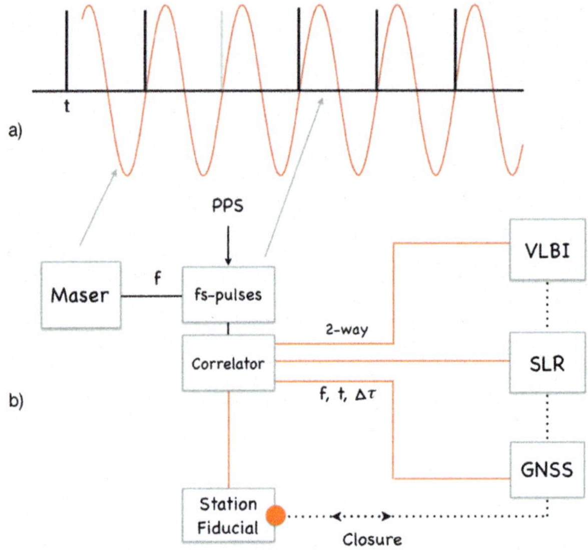

Fig. 6 The fs-mode-locked laser is locked to the stable reference frequency from the maser, so that the ultra-short laser pulses correspond to the zero-crossings of the maser frequency (**a**). In order to supply the measurement instrumentation the pulse train is split up in several fiber lines, each of which is stabilized by the correlation unit after a two-way passage (**b**). At the endpoints both, the reference frequency and the epoch can be recovered from the pulse train. The delay for each line is an integer multiple of the pulse spacing and is held constant by the feedback system of the correlator. This creates a coherent distribution of time across the observatory and provides the opportunity to apply the closure measurement technique for the reduction of systematic errors

structure is not lost due to dispersion in the fiber. At the end of the fiber a portion of the light is back-reflected towards the origin. At the input beam splitter it mixes with the pulses reflected from the short arm. The superimposed pulses are correlated and a feedback loop controls a delay line such, that the delay relationship is servoed to constant delay. This active control provides a long-term stable constant delay between both ends of the signal line. The short laser pulses generate timing signals with sub-ps resolution and from the repetition rate the clock frequency can be recovered.

Figure 6(a) sketches the full campus distribution. The pulse train from the mode-locked laser is split up in several synchronous channels, maintaining the coherence of the pulses. Each link is equipped with a balanced optical cross-correlator to compare the phase relationship of the outgoing and returning laser pulses as they are transmitted over a monomode optical fiber to the remote instrumentation (Kim et al. 2007). Each link in this star configuration is stabilized on an integer number of pulses with high accuracy. Therefore the delay introduced by each signal line is constant and known. Temperature effects on the fiber are compensated instantaneously by the feedback circuit. Since each time and frequency distribution line is independently controlled, also the delay between the various measurement systems of the observatory becomes both constant and known, thus providing the basic requirement for time transfer. The block diagram in Fig. 6(b) illustrates this setup. The VLBI, SLR and GNSS systems are all connected on individual fiber lines and obtain 4 different quantities from the time and frequency distribution unit, which are the equidistant time pips derived from the short laser pulses, the frequency of the reference oscillator contained in the repetition rate of the pulses, the beginning of a second, derived from one specifically

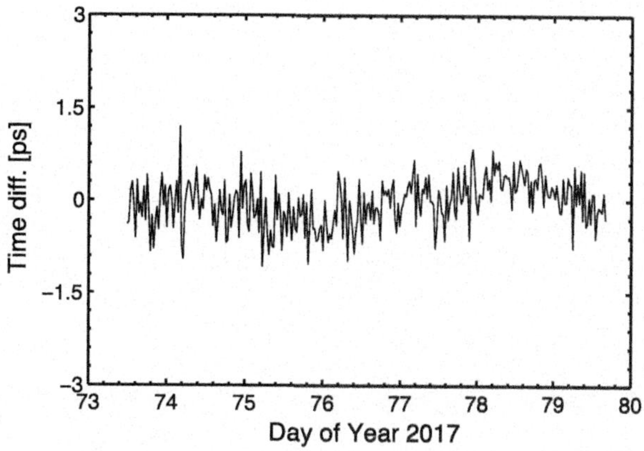

Fig. 7 Time comparison between two links of the time and frequency distribution system. Each link has a fiber length of more than 100 m and over about one week of operation the time delay between these two terminals remains stable to within about 1 ps

defined pulse per second out of the entire pulse train and the delay between the beginning of a new second at the central distribution unit and the accumulated delay from half the number of pulses of the roundtrip on the fiber line. Therefore it is possible to define a single timescale for the entire observatory, which is independent of any other measurement system and relates all the measurements across all space geodetic techniques to a single point of reference.

3.3 Delay Extraction by the Virtue of Closure Measurements

In order to define a unique reference point in time and space, we have created a local reference target for VLBI, SLR and in the future also GNSS. This geometrical reference point is also linked to the time and frequency distribution system by an independent fiber link. For SLR this means that it can be used for one-way ranging as well as for two-way ranging at the same time. In other words, it is possible to establish the accuracy of the SLR system delay by measuring the closure in time between the SLR clock and the target clock by transferring the time from the SLR system to the reference target. Any system delay that is not properly included in the SLR system calibration and that exceeds a value of a few ps, shows up as an offset between the two timing terminals and can be unambiguously identified and quantified. Furthermore it is possible to establish the long-term stability of time and frequency distribution system as well as the SLR systems on the observatory. In order to demonstrate the resolution and stability of the novel time and frequency distribution system, we have compared two independent links in different buildings more than 50 m apart with each other, using an asynchronous two-way time comparison system (Kodet et al. 2015). Figure 7 depicts the results. Over almost 7 days both terminals agreed to within 1 ps with each other. Owing to the active stabilization there is no temperature related drift left on these continuous time comparisons. It is important to note that the ultimate resolution for the transfer of time over these fiber links is depending on the pulsewidth of the train of pulses from the mode-locked clock laser and will be of the order of 10 fs for the optical part of the link. However, since the time is finally accessible as an electric square wave signal at both ends of the fiber line and these electrical signals are bandwidth limited to less than 8 GHz and in the presence of noise the practical performance limit for the resolution of the time comparison is currently at 1 ps.

Due to the controlled delay of the timing system of the observatory the laser oscillator and the VLBI source are operated phase coherently with respect to each other. This means that the intra-technique calibration must provide a closure with zero offset for a measurement process that starts from the common clock and leads back to the clock via the common reference target with all system delays under control. Once this is achieved, it is also possible to carry out inter-technique calibration by closure measurements from say the SLR system to the VLBI technique. Since VLBI source and SLR laser oscillator are synchronized and with constant and known delay relative to each other, erroneous system delays will become visible. Although the first several links for the delay compensated time and frequency distribution are already in place at the time of writing, the full system is far from being completed.

4 Optical Ground to Space Time Transfer

4.1 Einstein Synchronization

Giorgetta et al. have demonstrated that fs-mode-locked lasers can be used to transfer time over a fixed 2 km free space distance with a precision of 0.2 fs at 1 second of integration (Giorgetta et al. 2013). Deschenes et al. have taken this concept further to a 4 km terrestrial distance and they could show that the long-term stability of the time transfer reached a value of 40 fs peak to peak over two days in the presence of a turbulent atmosphere (Deschênes et al. 2016). We are setting up an all optical time transfer link between ground and the Atomic Clock Ensemble in Space (ACES) onboard the International Space Station (ISS) in space. Since the use of a fs-pulse laser is not directly applicable to the dynamic satellite orbits, we combine the fs-mode-locked laser pulses from the local timing system with the SLR technology. Due to its nature as an optical two-way time of flight measurement technique, Satellite Laser Ranging constitutes the only viable technical realization of an Einstein synchronization process (Einstein 1905). Therefore SLR provides the best suited approach for the accurate comparisons of clocks and time transfer even on intercontinental distances (Degnan 2002). In the idealized theoretical thought experiment of the Einstein publication, both the generation of the light signal as well as the detection of the presence of such a signal is instantaneous and there is no spatial offset between the light signal origin, the reflection and the clocks involved. In practice however, there are delays associated both with the generation and the detection of short laser pulses as outlined in the previous sections. There are two clocks, one of them, clock A is represented in the inertial frame of a local observatory and the other B is located in another inertial frame at some distance for example on an orbiting satellite.[1] A short laser pulse is generated by a mode-locked laser and passes through an optical beam splitter associated with the position of the clock A, detecting the outgoing (t_A) and incoming (t'_A) laser pulses for the synchronization process. We note that this detection process happens at an delayed instance, which is not a priori known. Figure 8, drawn in the form of a Michelson interferometer, illustrates this setup. In order to detect this transmitted light signal, a small portion of this laser pulse is passing through a reference arm of precisely known length, and after traversing the beam splitter it is detected and timed on clock A. The signal on the measurement branch propagates to the remote clock B and is back reflected by a cube corner reflector instantaneously, so that no extra delay is introduced. A small portion of the laser beam is used to detect and time tag

[1] For the simplicity of arguments, we assume that the motion of the satellite clock can be approximated as a clock in another inertial frame. At the current level of measurement resolution, we may ignore the presence of rotation.

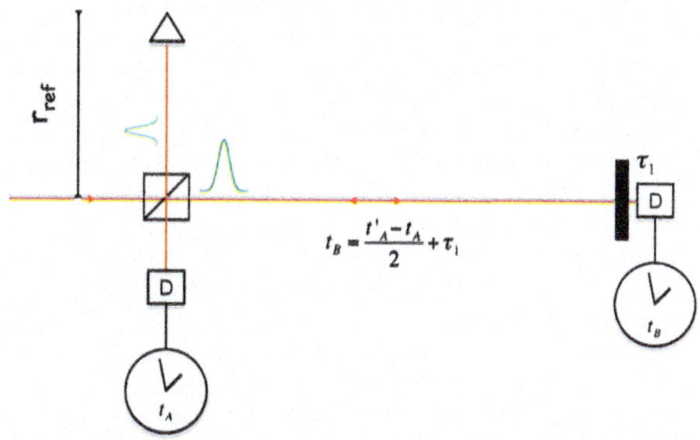

Fig. 8 Illustration of the Einstein synchronization, drawn like a Michelson Interferometer. Since neither the detection process itself nor the local signal path can be realized without delay, the integration of a reference arm is needed for technical reasons. The detector delay τ_1 cannot be compensated and has to be made as small and constant as possible

the arrival of the light signal at clock B. This yields the reading t_B. Again, this process is not instantaneous and the extra delay τ_1 accounts for that. Let us design the detector such that τ_1 is very small and that it can be established with very little uncertainty before launch. This is a reasonable assumption,[2] because the remote detector requires only very few electronic components and a short signal cable to connect up to the timer. The two clocks compare as

$$t_B = t_A + \frac{t'_A - t_A}{2} + \tau_1, \tag{2}$$

with the delay τ_1 stable and known. Although this concept of Einstein synchronization could in principle be applied between remote clocks on the ground, the really useful application is the clock synchronization between ground and space.

4.2 Time Transfer Between Ground and the Atomic Clock Ensemble in Space

One of the major objectives of the ESA project "Atomic Clock Ensemble in Space" (ACES) is the measurement of the gravitational redshift with unprecedented accuracy (Cacciapuoti and Salomon 2009). For this purpose there are two highly stable clocks provided; one is a laser cooled cold Cs-clock (primary standard) with a stability of the order of 3×10^{-16} at one day and an accuracy of 10^{-16}, the other is a hydrogen maser suitable for the application in space. It achieves 2×10^{-15} in 1000 seconds. ACES will be attached to the Columbus module of the International Space Station (Hess et al. 2011). A GNSS receiver provides the orbit, so that the exact flight path of the orbiting clock can be reconstructed. The comparison between clocks on the ground and the ACES payload is achieved by Two-Way-Satellite Time and Frequency Transfer (TWSTFT) in the microwave regime. Since optical signals are neither affected by variations in the total electron content in the ionosphere nor by the variable part of water vapor in the troposphere, laser time transfer has also been adopted for this project as an independent component for the clock comparison process. Apart from more resilience of this technique against systematic errors on the propagation path, the laser time transfer method also provides a clear geometrical reference both in space and on the ground for the clock comparison process. There were two important developments required to integrate the time transfer functionality into the SLR operations. These are:

- a broadband detector onboard the satellite with an acceptance angle for the field of view of 120° and a stable and well established internal signal delay;

[2]The discussion of Sect. 2 on variable delays due to intensity variations of the detected signal however still applies.

Fig. 9 Image of the engineering module of the ACES time transfer detector. To the left one can see the wide field of view. A baffle is used to reduce the impact of the Earth albedo from the nadir direction

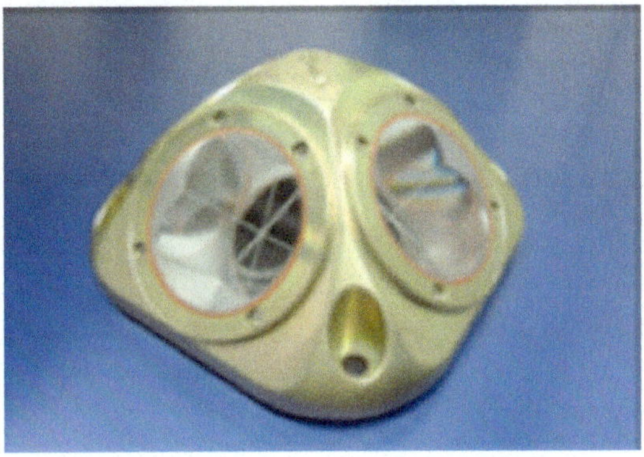

Fig. 10 The retro-reflector of the ELT experiment is based on the Champ/Grace design. Only a single cube corner is contributing to the back-scattered signal at any time

- a delay-compensated link to the local hydrogen maser, so that the SLR system is tightly referenced to the observatory clock and does not introduce extra delays, which would increase the error margin unduly.

In order to minimize the hardware requirements for the ACES payload, a solid state detector with high bandwidth and periodic self-gating was chosen. The pulse forming electronics was reduced to a minimum in order to reduce possible variations of the signal delay, which would otherwise be introduced by the detector. As a consequence, the complexity of the time transfer is shifted to the ground station, which has to fire the laser such, that the laser pulse reaches the spaceborne detector within 100–1000 ns after it is gated on. This gating process repeats with every 10 ms starting from the beginning of a second in the UTC timescale as it is realized aboard the ISS. In order to minimize any detector delay, the detector is capable of timing events from single photons. Therefore the scattered light from the Earth albedo is a significant noise source; in particular for a detector with such a wide acceptance angle. In Fig. 9 the engineering module of the ACES optical detector is shown. The event timing of the laser pulses at the satellite is carried out by a spare channel of the TWSTFT hardware of the ACES payload. The photo-detector is located near a small optical retro-reflector array, which is shown in Fig. 10. In order to align the retro-reflector position with the detector, the eccentricity between the detector and the line of sight of the respective SLR station and reflector has to be considered. Therefore it is necessary to keep track of the attitude of the Columbus module relative to the respective SLR observatory and apply the geometrical offsets as a correction. Attitude information of the ISS is available from the downlink metrology data of the space station. Expectations are, that the optical time transfer will provide accurate time transfer ground to space to within 30 ps on a single shot. At the time of writing the ACES payload is completed and awaits a suitable launch vehicle. The deployment of ACES is scheduled for 2018. In the meantime the ground segment is getting ready for the time transfer experiment. Since ELT is the first SLR operation, which involves the presence of astronauts near the target, particular safety precautions have to be met in order to guarantee that the maximum permissible energy level for pulsed laser light is not exceeded by any SLR station ranging to the ISS. The closure measurement between the SLR system and the central fiducial target of the observatory is carried out by a combination of one-way and two-way ranging. While the two-way roundtrip measurement is required to obtain the exact distance between the SLR system and the central target, the one-way range measurement is comparing the phase of the two linked clocks in order to establish a (variable) delay of the SLR system. For the establishment of the transmit system delay, the laser pulses fired from the SLR system are timed on a local detector at the central target. In a second step the fs-pulses of the timing system transmitted from the local target are detected and timed at the SLR system. This provides the receive system delay of the SLR system. Both measurements are combined to establish accurate agreement between the coherent clock signal at both locations. Consistence in these measurements and zero closure with respect to the survey of the target and the SLR station provide the desired zero offset for the time transfer between the ground clock and the ISS. The laser reflector shown in Fig. 10 is constructed in the way that only one reflector is contributing to the reflected signal. Therefore the target depth of this reflector array is comparable to a flat board. The transmit laser pulse width is as short as 10 ps (FWHM). This leaves intensity variations and the detector delay τ_1 (Eq. (2)) at the ISS as the main remaining contributor to the error budget of the time transfer. For the operation of the ELT time transfer it is mandatory to operate the laser at very low power, such that both for the echo detection on the ground and at the ISS the detectors work in the single photon detection regime. This minimizes the intensity depending error contribution and avoids systematic offsets between the two clocks for the comparison. The detector delay τ_1 is calibrated prior to the launch and turns out to be very stable with a value of 0.5 ps/K with respect to temperature changes and of the order of 12 ps (Prochazka et al. 2011, 2016). It is important to note that the variation of this predetermined value matters more than the absolute value of the delay itself. Finally the contribution of the orbit uncertainties and the attitude variation of the ISS need to be addressed. The location of the ACES clocks and the optical detector are tracked by a GNSS receiver located near the ACES clocks and the detector. From these measurements the clock motion along the orbit can be reconstructed and the necessary relativistic corrections applied. Taking all these error contributions together, it is expected that the time transfer will indeed be accurate to within 30 ps on a single shot.

4.3 T2L2 a Currently Existing Time Transfer Missions

While ELT is a project still under development, another time transfer experiment has already impressively demonstrated the feasibility of the laser time transfer concept at the level of a few pico-seconds after 30 seconds of averaging. *Time Transfer by Laser Link* (T2L2) is in orbit as a payload on the "Ocean Surface Topography Mission", JASON 2 satellite (Samain et al. 2009). Although JASON 2 only operates an ultra-stable-oscillator (USO) as the satellite clock, all essential functions for optical time transfer can be tested by this mission (Belli et al. 2016). At the satellite the detected laser pulses are time-tagged and transmitted to the ground via a telemetry link. A signal amplitude trigger threshold ensures, that very little noise events are recorded. The first step of the evaluation process uses the laser fire and

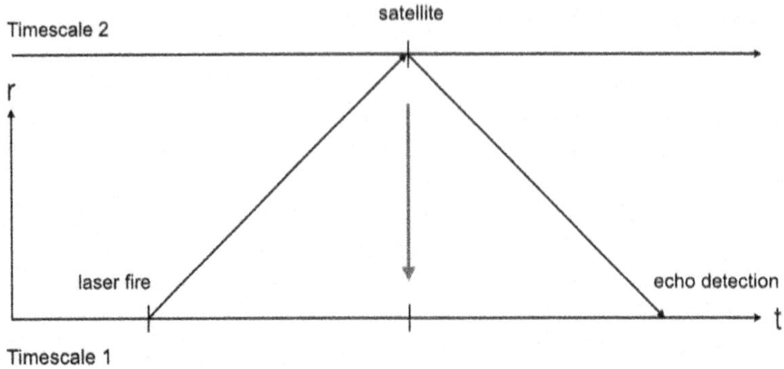

Fig. 11 Illustration of the time transfer process between a ground station and a satellite. The midpoint in time between laser fire and echo detection in timescale 1 corresponds to the arrival time of the laser pulse at the satellite detector

detection events to form data triplets, containing the start and stop event on the ground station along with the detections acquired by the satellite. For this purpose the offset between the timescale on the ground and on the satellite is usually known a priori with sufficient accuracy to allow the alignment of the measurements (Samain et al. 2014, 2015). Figure 11 illustrates this relationship in a range-time-diagram. When the laser is fired, the laser pulse moves away from the ground station as time progresses. When the light hits the cube corner and the detector, it propagates back to the ground station, where it is detected, when the roundtrip is completed. For satellites in a near Earth orbit, the uplink time is equivalent to the downlink time with good approximation. Therefore the epoch corresponding to the start epoch plus 1/2 of the measured roundtrip time interval in the timescale of the ground station (timescale 1) is equivalent to the detected epoch of the arrival time at the satellite in the timescale of the spacecraft, once the delay of the optical detector on the satellite is applied. Since the position of the detector and the retro-reflector on the satellite do not coincide, the respective eccentricity has to be corrected using the satellite attitude information, provided by the satellite mission. In this measurement process other systematic errors, such as for example an intensity dependent delay have either to be avoided or are required to be corrected before the respective timescales can be synchronized. For a passage of the JASON 2 satellite over the Geodetic observatory in Wettzell (Germany) the comparison of the USO oscillator on the satellite to the hydrogen maser of the observatory yielded a resolution for the time transfer of 30 ps for 1 second of averaging and 7 ps for 30 seconds of averaging (Samain et al. 2009). After 30 seconds of integration the drift of the USO became dominant. It is expected that the better ACES clocks will improve the resolution of optical time transfer further.

5 Summary

Small and variable system delays are currently the most obvious error sources in the techniques of space geodesy. They prohibit the use of time as an observable in space geodesy and are only visible as offsets in the measurements of the SLR, VLBI and GNSS technique. As long as time is not coherently distributed over an entire observing station, is it impossible to identify and remove the various contributors for variable system delays in the instrumentation. In the absence of any control mechanism, it must be suspected that the problem of small variable system delays is exacerbated by tiny asymmetries in the operation of the measurement systems and their calibration. Large and extended instrumentation like the VLBI and SLR hardware have many places where small system delays are potentially degrading the measurement quality at a subtle level. Section 2 discussed the obvious error sources in SLR. However, when the distribution of time and frequency is made coherent over the entire set of instrumentation, a novel way of delay control becomes available. Closure measurements, starting out from the coherent clock phase and passing through the entire instrumentation before coming back to the same clock in a different location, are powerful tools for systematic error identification. This however requires a stable time and frequency distribution system, where variations in the system delay are actively controlled and compensated or accurately measured. For larger observatories, where multiple techniques of space geodesy are present, the additional installation of a multi-technique ground target, connected to the station clock, offers a unique way of performing closure measurements within and across the various observation techniques. Such a target also constitutes a unique reference point for space and time, opening the door for using time as an observable in space geodesy, provided that a suitable process for clock synchronization over global distances can be established. For this purpose, optical time transfer via SLR will play an important role. In an environment, where major national standards laboratories currently develop highly accurate two-way frequency comparison techniques, in order to compare the frequency of optical clocks over distances of thousands of km (Lisdat et al. 2016), laser time transfer by SLR techniques can be a valuable extension on the route to extend the coherence of time over larger distances on the ground and also into the near Earth orbit. Precise clocks in space like those of the ACES project are very suitable candidates for this purpose.

Acknowledgements The authors acknowledge support from the Wettzell SLR team, namely Johann Eckl, Stefan Riepl and Günther Herold. This work was supported by the Deutsche Forschungsgemeinschaft contract SCHR 645/6-5 within the Research Group FOR1503 and by the Bundesamt für Kartographie und Geodäsie.

References

G. Appleby, J. Rodríguez, Z. Altamimi, Assessment of the accuracy of global geodetic satellite laser ranging observations and estimated impact on ITRF scale: estimation of systematic errors in LAGEOS observations 1993–2014. J. Geod. **90**(12), 1371–1388 (2016)

A. Belli, P. Exertier, E. Samain, C. Courde, F. Vernotte, C. Jayles, A. Auriol, Temperature, radiation and aging analysis of the DORIS Ultra Stable Oscillator by means of the time transfer by laser link experiment on Jason-2. Adv. Space Res. **58**(12), 2589–2600 (2016). https://doi.org/10.1016/j.asr.2015.11.025

J. Blazej, I. Prochazka, Avalanche dynamics in silicon avalanche single- and few-photon sensitive photodiode. J. Phys. Conf. Ser. **193**(1), 012041 (2009)

L. Cacciapuoti, C. Salomon, Space clocks and fundamental tests: the ACES experiment. Eur. Phys. J. Spec. Top. **172**(1), 57–68 (2009)

J.J. Degnan, Millimeter accuracy satellite laser ranging: a review, in *Contributions of Space Geodesy to Geodynamics: Technology*. AGU Geodynamics Series, vol. 25 (1993), pp. 133–161

J.J. Degnan, Asynchronous laser transponders for precise interplanetary ranging and time transfer. J. Geodyn. **34**(3–4), 551–594 (2002). https://doi.org/10.1016/s0264-3707(02)00044-3

J.-D. Deschênes, L.C. Sinclair, F.R. Giorgetta, W.C. Swann, E. Baumann, H. Bergeron, M. Cermak, I. Coddington, N.R. Newbury, Synchronization of distant optical clocks at the femtosecond level. Phys. Rev. X **6**(2), 5109–5115 (2016)

A. Einstein, Zur Elektrodynamik bewegter Körper. Ann. Phys. **217**, 891–921 (1905)

F.R. Giorgetta, W.C. Swann, L.C. Sinclair, E. Baumann, I. Coddington, N.R. Newbury, Optical two-way time and frequency transfer over free space. Nat. Photonics **7**(6), 434–438 (2013)

M.T. Hassan, T.T. Luu, A. Moulet, O. Raskazovskaya, P. Zhokhov, M. Garg, N. Karpowicz, A.M. Zheltikov, V. Pervak, F. Krausz, E. Goulielmakis, Optical attosecond pulses and tracking the nonlinear response of bound electrons. Nature **530**(7), 66–70 (2016)

M.P. Hess, L. Stringhetti, B. Hummelsberger, K. Hausner, R. Stalford, R. Nasca, L. Cacciapuoti, R. Much, S. Feltham, T. Vudali, B. Léger, F. Picard, D. Massonnet, P. Rochat, D. Goujon, W. Schäfer, P. Laurent, P. Lemonde, A. Clairon, P. Wolf, C. Salomon, I. Prochazka, K.U. Schreiber, O. Montenbruck, The ACES mission: system development and test status. Acta Astronaut. **69**(11), 929–938 (2011)

J. Kim, F. Ludwig, M. Felber, F.X. Kärtner, Long-term stable microwave signal extraction from mode-locked lasers. Opt. Express **15**(1), 8951–8959 (2007)

J. Kodet, P. Panek, I. Prochazka, Two-way time transfer via optical fiber providing subpicosecond precision and high temperature stability. Metrologia **53**(1), 18–26 (2015)

C. Lisdat, N. Quintin, C. Shi, S.M.F. Raupach, C. Grebing, D. Nicolodi, F. Stefani, A. Al-Masoudi, S. Dörscher, S. Häfner, J.-L. Robyr, N. Chiodo, S. Bilicki, E. Bookjans, A. Koczwara, S. Koke, A. Kuhl, F. Wiotte, F. Meynadier, E. Camisard, M. Abgrall, M. Lours, T. Legero, H. Schnatz, U. Sterr, H. Denker, C. Chardonnet, Y. Le Coq, G. Santarelli, A. Amy-Klein, R. Le Targat, J. Lodewyck, O. Lopez, G. Grosche, P.E. Pottie, A clock network for geodesy and fundamental science. Nat. Commun. **7**, 1–7 (2016)

I. Prochazka, J. Blazej, J. Kodet, Measurement of the optical to electrical detection delay in the detector for ground-to-space laser time transfer. Metrologia **48**(3), L13–L16 (2011)

I. Prochazka, J. Kodet, J. Blazej, Note: space qualified photon counting detector for laser time transfer with picosecond precision and stability. Rev. Sci. Instrum. **87**(5), 056102 (2016). https://doi.org/10.1063/1.4948733

E. Samain, P. Exertier, P. Guillemot, F. Pierron, D. Albanese, J. Paris, J.M. Torre, I. Petitbon, S. Leon, Time transfer by laser link T2L2 first results, in *The 2009 Joint Meeting of the European Frequency and Time Forum (EFTF) and the IEEE International Frequency Control Symposium (FCS)* (2009), pp. 194–198

E. Samain, P. Vrancken, P. Guillemot, P. Fridelance, P. Exertier, Time transfer by laser link (T2L2): characterization and calibration of the flight instrument. Metrologia **51**(5), 503 (2014)

E. Samain, P. Exertier, C. Courde, P. Fridelance, P. Guillemot, M. Laas-Bourez, J-M. Torre, Time transfer by laser link: a complete analysis of the uncertainty budget. Metrologia **52**(2), 423 (2015)

Space Sci Rev (2017) 212:1423–1432
DOI 10.1007/s11214-017-0419-8

Data Timing, Time Transfer and On-board Clock Monitoring for Space Astrometry with *Gaia*

Sergei A. Klioner[1] · Robin Geyer[1] · Hagen Steidelmüller[1] · Alexey G. Butkevich[1]

Received: 18 April 2017 / Accepted: 2 September 2017 / Published online: 2 October 2017

Abstract The paper discusses the problematic of data timing in the framework of the space astrometry mission *Gaia*. For various reasons related both to astrometry and to the measuring principles of the *Gaia* instrument it is mandatory to assign a highly stable and accurate time tag for each observational point. For this purpose *Gaia* has a Rb clock on board. That on-board clock is a free-running oscillator and must be regularly synchronized with TCB which is used as the underlying relativistic coordinate time scale in the whole *Gaia* data processing. To monitor the reading of the on-board clock with respect to TCB (or any other timescale) a one-way clock synchronization scheme is implemented. This scheme takes into account all known theoretical effects (e.g., relativity, tropospheric delay, etc.) and allows one both to monitor the health of the on-board clock and to create a clock model at the accuracy of better than 1 microsecond.

Keywords Gaia · Astrometry · Space mission · Time synchronization · Atomic clocks in space

1 The *Gaia* Mission

ESA's second space astrometry mission *Gaia* was launched in December 2013 and after an extended commissioning period started its scientific operations in July 2014. During its routine science operations *Gaia* has already delivered an immense data set of high-accuracy positional observations. The details of the mission and the principles of measurements can be found in (Prusti et al. 2016). In spite of some unexpected difficulties with the instrument, *Gaia* Data Processing and Analysis Consortium (DPAC) published the *Gaia* Data Release 1

High Performance Clocks with Special Emphasis on Geodesy and Geophysics and Applications to Other Bodies of the Solar System
Edited by Rafael Rodrigo, Véronique Dehant, Leonid Gurvits, Michael Kramer, Ryan Park, Peter Wolf and John Zarnecki

✉ S.A. Klioner
Sergei.Klioner@tu-dresden.de

[1] Lohrmann-Observatorium, Technische Universität Dresden, 01062 Dresden, Germany

(Brown et al. 2016; Lindegren et al. 2016) in September 2016 and is working towards the *Gaia* Data Release 2 in 2018.

Gaia is a very complicated instrument specially designed for an automatic high-accuracy astrometric survey in space. *Gaia* has two almost identical astrometric telescopes sharing the same focal plane that represents the largest CCD camera ever flown in space. The focal plane consists of 106 CCDs of a total of about 1 gigapixel. To reduce all possible optical aberrations and instabilities, the instrument has no moving parts and is build of a highly stable material—SiC. Two telescopes observing simultaneously and in substantially different directions (106.5° in the case of *Gaia*) are one of the key features of the design and allow one to measure absolute parallaxes. Except for some periods of maintenance, *Gaia* is designed to observe permanently. Satellite is forced to follow a special smooth observational schedule—called the Scanning Law—that allows one to achieve a sky coverage that is as homogeneous as possible and also maximize the accuracy of astrometric results. Both the Scanning Law and the orbit of the satellite (a Lissajous orbit around the Lagrange point L_2 in about 1.5×10^8 km from the Earth) are designed to ensure high thermal and mechanical stability of the instrument. *Gaia* automatically recognizes all celestial sources appearing in the field of view, automatically distinguish between real sources and signals e.g. from cosmic rays, and sends small set of pixels (called "window") around each recognized source to the ground. *Gaia* is designed to observe up to several hundred thousands of sources simultaneously and has a powerful and sophisticated computing and controlling equipment on board. *Gaia* ground processing is aimed to derive astrometric and photometric information from this data. The astrometric accuracy of *Gaia* depends on the brightness of the source and to a lesser degree on the position on the sky (due to some remaining inhomogeneity of the sky coverage). For the optimal stars, the accuracy at the end of the mission will reach a level below 10 microarcseconds. The data processing is very complicated and represents one of the most sophisticated and massive computations in the history of astronomy: from about 1 Petabyte of data comprising approximately 10^{12} individual observations one should derive up to 10^{10} parameters characterizing more than 10^9 celestial objects. Further details of both the instrument and the data processing can be found in the publications cited above.

One particular aspect of the data processing is related to the fact that each *Gaia* observation must be tagged with the corresponding time of observation. This paper discusses this particular aspect of the *Gaia* data processing. The time tags should be both stable and accurate (see below). To ensure this *Gaia* has got a Rb clock on board. The time tags are essential part of the *Gaia* measurement concept. If those time tags got corrupted, the *Gaia* data would become practically useless. The Rb clock of *Gaia* is a free running oscillator that should be monitored, related to and synchronized with widely-used time scales like UTC or TCB, realization of which is based on ensembles of high-accuracy clocks on the Earth. This is done using specially organized one-way synchronization observations between ESA ground stations and *Gaia* satellite. Because of the high accuracy of *Gaia* observations and products, the observational data of *Gaia* must be interpreted and modeled in a consistent relativistic framework. This is also true for the *Gaia* on-board clock monitoring.

In this paper we will use standard notations of the time scales and astronomical reference frames like UTC, TCB, ITRF etc. Readers not familiar with these notations are invited to consult e.g. Soffel et al. (2003), Seidelmann and Urban (2013).

2 Requirements for the Timing of *Gaia* Data

Each observation of *Gaia* must be tagged with the reading of the *Gaia* clock latched at some fiducial moment within that individual observation. A free-running Rb oscillator installed on board of *Gaia* produces time tags that are called On-Board Time (OBT). OBT is a technical time which reflects all the imperfections of the particular clock. OBT can even have jumps related to e.g. possible resets of the clock. OBT is a purely technical time with no a priori relation to any other time scales. *Gaia* data processing uses TCB as the underlying time scale. TCB is realized by ensembles of Earth-based clocks via TAI or TT. TCB is used e.g. to interrogate solar system ephemerides or to link *Gaia* observations with observations performed by other instruments. Clearly, the OBT must be converted to TCB using some synchronization procedure.

Timing requirements for *Gaia* consist of two parts. As usual in the discipline of time, one should carefully distinguish between *stability* and *accuracy* of the timing information. Some of the digital hardware components on board of *Gaia* (e.g., the CCDs) must be driven by some frequency standard. Space missions often use quartz (crystal) oscillators for this purpose. The stability of a modern space-qualified crystal oscillator can achieve 10^{-9} in a thermally stable environment of *Gaia*, but it is not sufficient for *Gaia*. Indeed, the Scanning Law of *Gaia* implies a rotation with an angular velocity of $60''$/sec or 0.6 microarcsecond (μas) in 10 nanoseconds. The ultimate accuracy of centroiding for one observation is expected to be at the order of 10 μas. Systematic errors over periods of time shorter than the rotational periods are very dangerous for the quality of astrometric products of *Gaia*. Such short-period systematic errors are extremely difficult and in certain cases impossible to calibrate. For this reason, it was decided that no systematic errors over the rotational period of 6 hours or 21600 sec should be introduced by the time tags of the data. This means that the timing error should be below 10 nanoseconds over the period of 21600 seconds, which implies the frequency stability of the on-board clock below 10^{-12}. For this reason *Gaia* got an atomic (Rb) clock with the stability reaching $\lesssim 10^{-13}$ over the *Gaia* rotational period of 21600 seconds.

The accuracy requirements for the OBT tags are very different in their nature and origin. A constant shift in the *Gaia* clock phase and/or in its frequency would be no problem in a purely *Gaia*-internal data processing provided that the frequency stability remains at the level discussed above. The relation between the OBT readings and some *Gaia*-external time scales becomes important as soon as some auxiliary data are used in the data processing. Those auxiliary data (e.g. *Gaia* or solar system ephemerides) are usually parametrized using some TAI/TT-based time scale (TT, TCG, TCB, etc.). The relation between OBT and, say, TCB is needed to interrogate the auxiliary data with the time argument of certain accuracy. Similarly, analyzing some time-dependent phenomena or predicting some astronomical events one wishes to express the analysis in terms of generally available time scales. One can think of several different sources of requirements coming from different types of time-dependent astronomical phenomena: variability of stars, motion of binary stars, motion of asteroids, etc. Taking into account all these applications, the official goal for the timing accuracy for *Gaia* was taken to be 1.7 microseconds. This accuracy represents also a limit of accuracy achievable without any substantial upgrade of the timing hardware available at the ESA ground stations.

The on-board clock of *Gaia* is the space-qualified Rubidium Atomic Frequency Standard (RAFS) produced by Spectratime: http://www.spectratime.com/products/ispace/rb/RAFS/. This is the same kind of clock as used e.g. by the Galileo GNSS program. The characteristics of the clock can be easily found in the literature (see e.g. Rochat et al. 2005). In particular, the RAFS has stability of 4×10^{-14} over 10000 sec and frequency accuracy of 2×10^{-10} with a drift of up to 2×10^{-10} per year. Before launch, the RAFS of *Gaia* was tested in a laboratory over a long interval of time. Those tests basically confirmed the specifications (see Fig. 1).

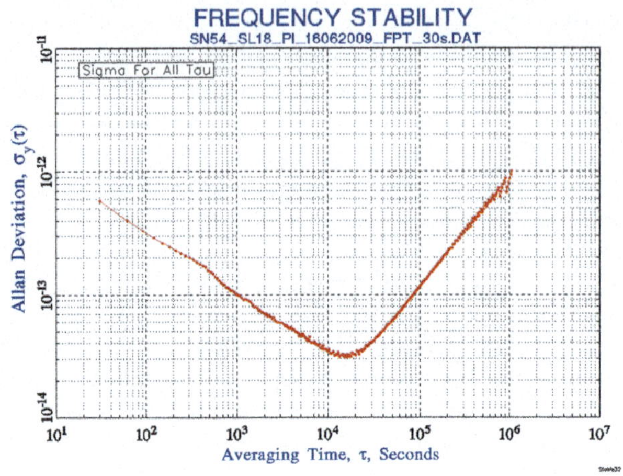

Fig. 1 Frequency stability of the *Gaia* RAFS represented by its Allan deviation (Allan 1966) from a test campaign before launch

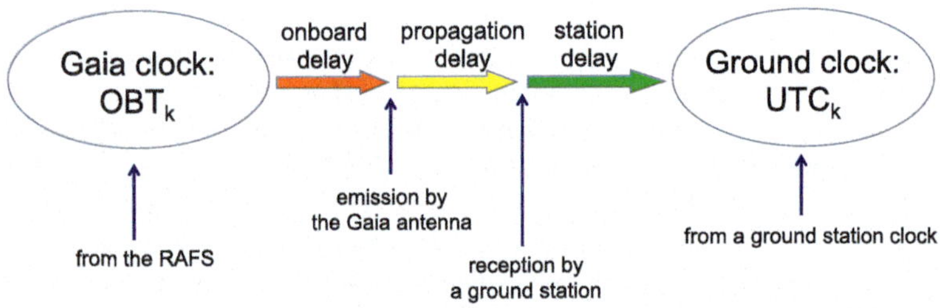

Fig. 2 Schematic view of the one-way clock synchronization

3 One-Way Clock Synchronization Process

There are many technical ways to synchronize remote clocks (see, e.g. Klioner 1992). In principle, for a microsecond accuracy with a pair of clocks separated by 1.5 million kilometers, the so-called two-way clock synchronization is the method of choice. Two-way approach is a technical realization of the Einstein clock synchronization procedure corrected for relative motion of the clocks and for certain relativistic effects. In the two-way approach, a special signal generated by a reference station is sent to the remote clock, where it is received and, after some well-calibrated delay, transmitted back to the reference station together with the time reading of the remote clock, at which the signal was received. The reference station delivers both the time of emission and reception according to its own reference clock. This information is sufficient to synchronize the remote and reference clocks in the sense of relativistic coordinate synchronization. The well-known advantage of this method is that the requirements for the accuracy of position of the remote clock can be substantially relaxed. This is also true for the tropospheric delay: the delays on the way from the reference station to the remote clock and back almost cancel out.

However, because of the technical difficulties for the *Gaia* design and considering the hardware readily available on the ESA Tracking Stations (ESTRACK), the two-way clock synchronization was dropped early at the design stage of *Gaia*. A straightforward one-way clock synchronization scheme was implemented instead. In the case of *Gaia* this one-way clock synchronization scheme is depicted on Fig. 2 and can be summarized as follows:

(1) a signal is generated on board of *Gaia* that contains the momentary (latched) OBT value—we denote it as OBT_k—and initiates the process of transmission of a special data packet, called "time packet" to the ground station;
(2) after some on-board delay the time packet is emitted by *Gaia* to the ground station;
(3) after a propagation delay (that includes also the tropospheric delay) the time packet arrives to the phase center of the ESTRACK station antenna;
(4) after some ground-station delay the content of the time packet (in particular, the value of OBT_k) is registered by the ESTRACK station hardware by assigning the corresponding value of UTC according to the local clock—we denote it as UTC_k—and storing the time couple $(\mathrm{OBT}_k, \mathrm{UTC}_k)$ in a database.

Thus the raw data for the clock synchronization are the time couples $(\mathrm{OBT}_k, \mathrm{UTC}_k)$. These time couples are obtained at irregular intervals of time with a typical interval of 1.5 sec between the subsequent time couples during the visibility periods of *Gaia* (the periods of time during which *Gaia* can communicate with a ground station). Duration of the visibility periods varies depending on the data volume that should to be transmitted to the ground. Outside the visibility periods the transformation between OBT and TCB must rely on interpolation.

4 High-Accuracy Time Transfer Model

The accuracy of about 1 microsecond needed for *Gaia* implies a rigorous relativistic model for the one-way clock synchronization data described above. The model uses the relativistic reference systems BCRS and GCRS as a generic tool of high-accuracy modeling of astronomical data (Soffel et al. 2003). The model explicitly introduces the relativistic proper time of *Gaia*, which is denoted as TG and represents an ideal clock located at the *Gaia* center of mass. The displacement of the *Gaia*'s atomic clock from the *Gaia* center of mass is fully negligible in the current context. The deviation between OBT and TG solely reflects the errors of the *Gaia* clock (ignoring possible constant phase shift).

TG can be computed as function of TCB using the ephemerides of *Gaia* and the major solar system bodies—the Sun, major planets and the Moon. TG as function of TCB and TCB as function of TG are be defined by ordinary differential equations with the initial condition that TG=TCB at JD2457023.5 TCB (2015 January 01 00:00:00.0 TCB). As described in (Klioner 2015) these equations are integrated numerically using ephemerides of *Gaia*, the Sun, the Moon and the major planets of the solar system. As a solar system ephemeris a TCB-based ephemeris INPOP10e specially prepared for *Gaia* is used. *Gaia* ephemeris is updated once per week by the flight dynamics team of ESOC. The TG–TCB transformation is updated as soon as a new version of *Gaia* ephemeris becomes available. All time transformations relevant for *Gaia* data processing and the details of their implementation are summarized in (Klioner 2010, 2015).

The scheme of the overall model for the time couples $(\mathrm{OBT}_k, \mathrm{UTC}_k)$ is shown on Fig. 3. The idea of the model is to compute the value TG_k that correspond to each OBT_k, so that a model for OBT as function of TG can be constructed. To this end, one needs to compute the couples of OBT_k and TG_k corresponding to one and the same physical event. Various physical events can be chosen, but the most natural choice seems to be the event of emission of the time couple signal by *Gaia*.

On the one hand, OBT_k can be corrected for the on-board delay Δ_{ob} (as calibrated before launch) to get $\mathrm{OBT}_k^{\mathrm{em}} = \mathrm{OBT}_k + \Delta_{\mathrm{ob}}$. From now on superscripts "em", "rcp", "gs", "ob"

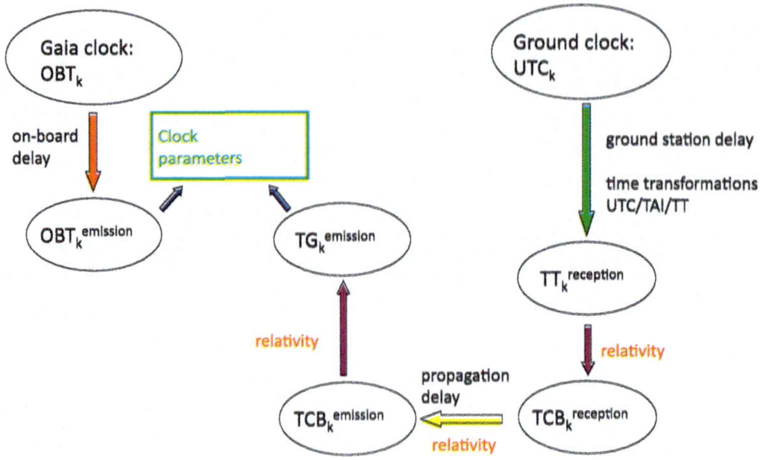

Fig. 3 The modeling scheme (see text for further details)

denote "emission", "reception", "ground station" and "on-board", respectively. On the other hand, the corresponding $\mathrm{TG}_k^{\mathrm{em}}$ is computed from UTC_k in a number of steps:

(a) The ground-station delay Δ_{gs} is subtracted from UTC_k to get $\mathrm{UTC}_k^{\mathrm{rcp}} = \mathrm{UTC}_k - \Delta_{\mathrm{gs}}$. The ground-station delay Δ_{gs} depends on the particular telemetry mode used at the moment of observation and is different for different ground stations.

(b) The moment of reception in TCB, $\mathrm{TCB}_k^{\mathrm{rcp}}$, is computed using a series of relativistic time transformations (Klioner 2010):

$$\mathrm{UTC}_k^{\mathrm{rcp}} \longrightarrow \mathrm{TAI}_k^{\mathrm{rcp}} \longrightarrow \mathrm{TT}_k^{\mathrm{rcp}} \longrightarrow \mathrm{TDB}_k^{\mathrm{rcp}} \longrightarrow \mathrm{TCB}_k^{\mathrm{rcp}}$$

Note that the transformation between TT and TDB is 4-dimensional and requires the position of the ground station in BCRS (this implies also the use of the Earth orientation in space).

(c) The TCB moment of reception $\mathrm{TCB}_k^{\mathrm{rcp}}$ is used to compute the TCB moment of emission of the corresponding signal by *Gaia* $\mathrm{TCB}_k^{\mathrm{em}}$. Here one should account for the tropospheric delay Δ_{tropo} and compute the TCB time interval that the signal needs to propagate from *Gaia* to the ground station in BCRS as discussed e.g. in (Klioner 2015). Here we consider the geometric delay and the relativistic Shapiro delay taking into account the motion of the ground station and the *Gaia* satellite.

(d) The TCB moment of emission $\mathrm{TCB}_k^{\mathrm{em}}$ is transformed into the corresponding moment of the *Gaia* proper time TG_k using the function TG(TCB) obtained as discussed above.

To complete this computation one needs a number of auxiliary data:

(i) the coordinates of all ESTRACK stations in the International Terrestrial Reference Frame 2008 (ITRF2008); this includes the Cartesian coordinates as well as the height above the geoid;

(ii) the coefficients of the mapping function for the tropospheric delay (the dry and wet Niell mapping functions are used (Niell 1996); a total of 9 coefficients for each ESTRACK station);

(iii) the adjustment of the dry delay due to the altitude difference of the barometer and the intersection of the azimuth and elevation axis of the antenna (used to compute the zenith delay from the Saastamoinen model (Saastamoinen 1972));

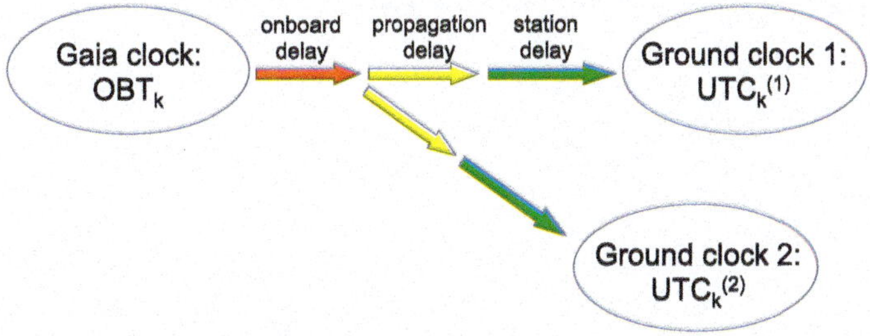

Fig. 4 Schematic view of the two-station experiment

(iv) the ground-station delays Δ_{gs} for each telemetry mode;
(v) the on-board delays Δ_{ob} for each telemetry mode;
(vi) *Gaia* ephemeris;
(vii) meteorological data—temperature, pressure and humidity—at the relevant ESTRACK station sampled once per minute (used to compute the dry and wet tropospheric delays at zenith);
(viii) information on the telemetry modes used at the time of transmission of each time couple;
(ix) the Earth orientation parameters (EOP) from the IERS, e.g. the EOP 08 C04 series available at https://www.iers.org/IERS/EN/DataProducts/EarthOrientationData/eop.html; the EOPs are used to compute the BCRS positions of the ground stations.

All these data were either provided before launch or are being provided or updated on a regular basis.

The model of the time transfer data results in a series of reduced time couples $(\mathrm{OBT}_k, \mathrm{TG}_k)$ that allow one to construct a relation between OBT and TG. As it was already pointed out, this relation reflects both the constant phase offset between OBT and TG (actual synchronization parameter that would persist even for ideal *Gaia* clock) and the imperfections of the *Gaia* clock. Since the *Gaia* clock is of high quality one can expect a simple relation between OBT and TG. On the other hand, the OBT–TG relation represents physical model of the *Gaia* clock as if the clock would be monitored in a laboratory. In this way the health status of the clock and its performance can be directly assessed.

The overall modeling accuracy of this algorithm is about 30 nanoseconds and is limited by the error of the distance between *Gaia* and the ground stations. Fortunately, since the *Gaia* satellite is observed by Doppler and radar techniques, the uncertainty in geocentric radial distance does not exceed 6–10 m. This accuracy is more than enough for the one-way clock synchronization with an accuracy of better than 1 microsecond.

5 The Two-Station Experiment

In order to test the validity and accuracy of the models and data used for the high-accuracy time synchronization, a special "two-station" experiment was performed: during a period of about two hours on 25 Feb 2014 the same time packets emitted by *Gaia* were independently received and registered by two ground stations (Fig. 4). In this way two sets of time couples were created: $(\mathrm{OBT}_k, \mathrm{UTC}_k^{(1)})$ and $(\mathrm{OBT}_k, \mathrm{UTC}_k^{(2)})$. Each set consists of 2994 time couples. These two sets of time couples were independently processed by the high-accuracy algo-

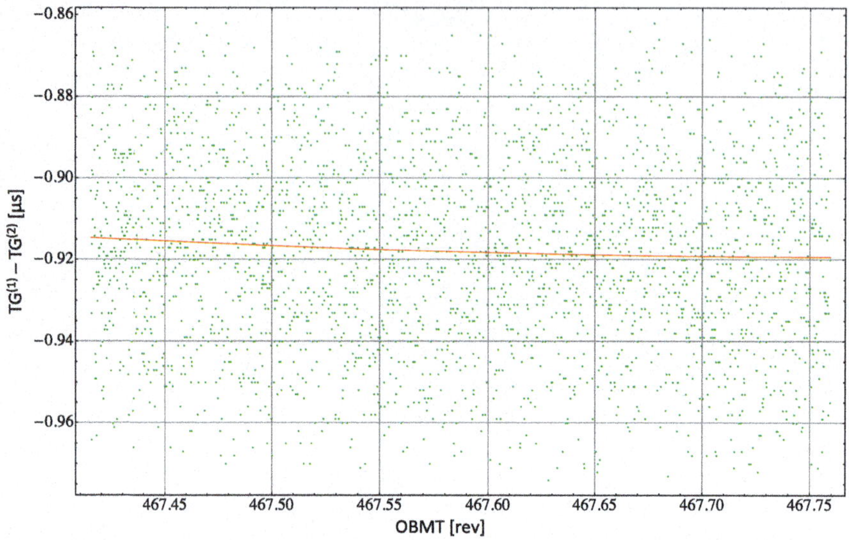

Fig. 5 The differences $TG_k^{(1)} - TG_k^{(2)}$ of the two-station experiment together with the fitted quadratic polynomial (*red curve*). The *x-axis* shows the corresponding OBT values in the units of "revolutions" roughly corresponding to 6 hours

rithm to obtain the corresponding $TG_k^{(1)}$ and $TG_k^{(2)}$. The differences $TG_k^{(1)} - TG_k^{(2)}$ are not influenced by the *Gaia* clock and should be zero in ideal case. They therefore characterize the errors of the model and/or data. These differences together with the quadratic polynomial fitted to the data are shown on Fig. 5. The polynomial fit shows the following.

- There was a desynchronization of -918 ± 1 nanoseconds (ns) between the clocks at two participating ground stations at the time of observation.
- Two clocks at the ground stations show apparent frequency offset of $(-6.7 \pm 2.0) \times 10^{-13}$. This frequency offset is too big to be attributed to the real ground station clocks and probably comes also from the imperfection of the tropospheric delay model. Note that the frequency drift was determined over the period of 2 hours only and is based on the change of the timing by about 5 ns over 2 hours. This can be considered as an ultimate limit for the accuracy of frequency transfer (syntonization) using the one-way time synchronization as implemented in *Gaia*.

After subtracting the fitted quadratic polynomial from the differences one gets the residuals of the model shown on Fig. 6. As one can see at the right side of the Fig. 6, the residuals have a zero-centered triangular distribution between approximately −57 ns and +57 ns. This should be expected because of the discretization error of the ground station clocks running at 17.5 MHz. This also demonstrates that the model described above works well and the remaining systematics is at the level below 10 ns. More details on the analysis of the two-station experiment can be found in (Klioner and Butkevich 2014).

We note here that this sort of experiments can be used to synchronize clocks on the ground stations. Other satellites with an orbit known to several ten meters can be used (the effect of the error of satellite position is much reduced in the differences $TG_k^{(1)} - TG_k^{(2)}$). Systematic errors in the synchronization using this approach can reach up to 10 ns mainly due to imperfections in the model of the tropospheric delay.

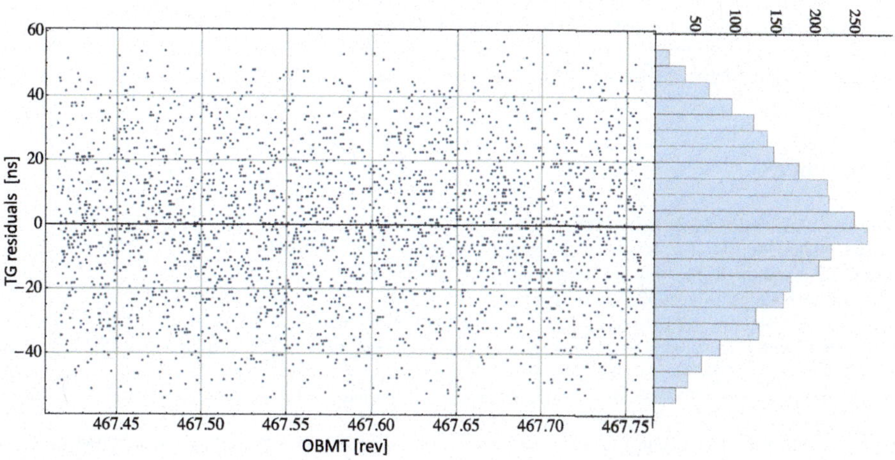

Fig. 6 Residuals of the two-station experiment: the triangular distribution of the residuals demonstrates that the model works as expected (see text). The *x-axis* shows the corresponding OBT values in the units of "revolutions" roughly corresponding to 6 hours

6 Timing Needs for a Hypothetical Space Mission for Nanoarcsecond Astrometry

For the accuracy level of *Gaia*, a clock with stability of about 10^{-13} over 6 hours is necessary and sufficient. Modern space-qualified Rb clocks perfectly fit this purpose. Nothing could be gained with *Gaia* data if a much more stable and/or accurate clock is used on board. It is interesting to check how the requirements would change if a hypothetical nanoarcsecond-level astrometric mission is considered.

Currently, it is not clear if nanoarcsecond astrometry is at all possible. Both technical difficulties and possible absence of sufficiently compact and bright sources in the Universe can impede the growth of astrometric accuracy. It may well be possible that astrometry beyond the level of about 0.1 μas is not possible. In the decades to come quite some efforts are planned to investigate the future of astrometry. Leaving aside all these problems and assuming the same Gaia-like measuring principles, one can analyze the accuracy requirement for the data timing for a nanoarcsecond astrometry mission.

Following the discussion of the timing requirements for *Gaia* and assuming that the same principles can be applied for a nanoarcsecond astrometry mission, one concludes that one needs a clock with the stability of 10^{-16}–10^{-17} over several hours and the time transfer between the satellite and ground stations with an accuracy of about 1 nanosecond. Clearly, a two-way link will be necessary for such a mission as well as something like a space-qualified optical clock. The time horizon for such a mission is however unclear. Obviously, at the moment these requirements are speculative.

Acknowledgements This work has made use of data from the European Space Agency (ESA) mission *Gaia* (https://www.cosmos.esa.int/gaia), processed by the *Gaia* Data Processing and Analysis Consortium (DPAC, https://www.cosmos.esa.int/web/gaia/dpac/consortium). Funding for the DPAC has been provided by national institutions, in particular the institutions participating in the *Gaia* Multilateral Agreement. The practical realization of the theoretical scheme for the time synchronization in Gaia is the work, which was prepared, discussed, organized and made possible by a large number of people from ESA and the DPAC. The help and fruitful discussion with Ulrich Bastian, Neil Cheek, Peter Collins, Sebastian Els, Rocio Guerra, Jose

Hernandez, Arek Kowalczyk, David Milligan, Hassan Siddiqui, and Gary Whitehead are gratefully acknowledged. The work was partially supported by the BMWi grant 50QG1402 awarded by the Deutsche Zentrum für Luft- und Raumfahrt e.V. (DLR).

References

D. Allan, Proc. IEEE **54**(2), 221 (1966)

Gaia Collaboration, A.G.A. Brown et al., Astron. Astrophys. **595**, A2 (2016)

Gaia Collaboration, T. Prusti et al., Astron. Astrophys. **595**, A1 (2016)

S.A. Klioner, Celest. Mech. **53**, 81 (1992)

S.A. Klioner, Relativistic time scales and time transformations for Gaia, Gaia internal report GAIA-CA-TN-LO-SK-01 available from the Gaia Livelink document archive (2010). http://www.rssd.esa.int/cs/livelink/open/3042112

S.A. Klioner, in *Recent Developments and Prospects in Ground-Based and Space Astrometry*, ed. by N. Capitaine, Z. Malkin (Paris Observatory, Paris, 2015), p. 55

S.A. Klioner, A.G. Butkevich, Results of the two-station timing experiment of 25 February 2014, Gaia internal report GAIA-C3-TN-LO-SK-019 (2014). Available from the Gaia document archive http://www.rssd.esa.int/cs/livelink/open/3263755

L. Lindegren et al., Astron. Astrophys. **595**, A2 (2016)

A.E. Niell, J. Geophys. Res., Solid Earth **101**(B2), 3227 (1996)

P. Rochat, F. Droz, P. Mosset et al., in *36th Annual Precise Time and Time Interval (PTTI) Systems and Applications Meeting* (2005), p. 26

J. Saastamoinen, *The Use of Artificial Satellites for Geodesy*, vol. 15 (1972), p. 247

P.K. Seidelmann, S.E. Urban (eds.), *Explanatory Supplement to the Astronomical Almanac* (University Science Books, Mill Valley, 2013)

M. Soffel et al., Astron. J. **126**, 2687 (2003)

Space Sci Rev (2018) 214:30
https://doi.org/10.1007/s11214-017-0459-0

Pulsar Timing and Its Application for Navigation and Gravitational Wave Detection

Werner Becker[1] · **Michael Kramer**[2] · **Alberto Sesana**[3]

Received: 28 May 2017 / Accepted: 11 December 2017 / Published online: 17 January 2018

Abstract Pulsars are natural cosmic clocks. On long timescales they rival the precision of terrestrial atomic clocks. Using a technique called pulsar timing, the exact measurement of pulse arrival times allows a number of applications, ranging from testing theories of gravity to detecting gravitational waves. Also an external reference system suitable for autonomous space navigation can be defined by pulsars, using them as natural navigation beacons, not unlike the use of GPS satellites for navigation on Earth. By comparing pulse arrival times measured on-board a spacecraft with predicted pulse arrivals at a reference location (e.g. the solar system barycenter), the spacecraft position can be determined autonomously and with high accuracy everywhere in the solar system and beyond. We describe the unique properties of pulsars that suggest that such a navigation system will certainly have its application in future astronautics. We also describe the on-going experiments to use the clock-like nature of pulsars to "construct" a galactic-sized gravitational wave detector for low-frequency ($f_{GW} \sim 10^{-9}$–10^{-7} Hz) gravitational waves. We present the current status and provide an outlook for the future.

Keywords Pulsars · Neutron stars · High precision timing · Atomic clocks · Pulsar timing · XNAV · Pulsar-navigation · Pulsar-based-navigation · Autonomous spacecraft navigation · Gravitational waves · Gravitational wave detector

---

High Performance Clocks with Special Emphasis on Geodesy and Geophysics and Applications to Other Bodies of the Solar System
Edited by Rafael Rodrigo, Véronique Dehant, Leonid Gurvits, Michael Kramer, Ryan Park, Peter Wolf and John Zarnecki

---

✉ W. Becker
web@mpe.mpg.de

M. Kramer
mkramer@mpifr-bonn.mpg.de

A. Sesana
asesana@star.sr.bham.ac.uk

[1] Max Planck Institute for Extraterrestrial Physics, Garching, Germany

[2] Max Planck Institute for Radio Astronomy, Bonn, Germany

[3] University of Birmingham, Birmingham, UK

1 Introduction

For millenia, keeping time was safely in the hands of astronomers, who watched the heavens to serve the societies' needs for time measurements. They used the Earth's rotation, the moon, the Sun and the stars to keep time, but, of course, they also exploited their observations to study and explore the Universe that we live in. Practical purposes, especially the need to navigate, turned the art of timekeeping into the hands of clockmakers and, hence, engineers and physicists. Today, world time is kept by a set of ultra-precise atomic clocks, and the sophistication and accuracy of these clocks is hugely impressive (see other contributions to this Topical Collection). The stability of these clocks is best on small timescales. When time stability is needed on long timescales, one can (and does) "hand over" time from clock to clock—or one can revert back to astronomical observations. Moreover, the fact that the Earth is moving on an elliptical orbit around the Sun, and hence at varying distances, means that every clock on Earth experiences a varying gravitational potential throughout the year. This leads to seasonal changes in clock rates that affect all clocks on Earth simultaneously. Astronomy can, hence, provide an "independent" time standard that is unaffected by effects on Earth or in the solar system. The key in providing such an astronomical time standard are objects called "pulsars.". Pulsars are compact, highly-magnetized rotating neutron stars which act as "cosmic lighthouses" as they rotate, enabling a number of applications as precision tools.

This contribution describes pulsars, the technique of pulsar timing and some of the resulting applications. Coming full circle, these applications include navigation and also time keeping. Here we concentrate on the former (see Sect. 2) and applications in fundamental physics, especially the detection of low-frequency gravitational waves (see Sect. 3). We note, that at the end, despite attractive features of a pulsar timescale, pulsars will not be able to compete with the precision and practicality of the best atomic clocks on Earth. Nevertheless, it is the combination of pulsar clocks with terrestrial clocks that allow, via clock comparison experiments, to probe a wide range of physics—and the Universe that we live in, continuing the tradition of astronomers.

1.1 Pulsars

Pulsars are born in supernova explosions of massive stars, created in the collapse of the progenitors core. Unlike most other astrophysical objects, pulsars emit across the whole electromagnetic spectrum (from radio to optical, X- and gamma-rays) at the expense of their rotational energy, i.e., the pulsar spins down as rotational energy is radiated away by its co-rotating magnetic field, a plasma wind, and broadband electromagnetic radiation. Thereby all radiation is powered by rotational energy, which distinguishes pulsars from "accretion powered" neutron stars. With the magnetic axis being inclined to the rotation axis, the pulsar acts like a cosmic light-house emitting a radio pulse that can be detected once per rotation when the beam is directed towards Earth (cf. Fig. 1).

The loss in rotational energy leads to an increase in rotation period, P, described by a measured $\dot{P} > 0$. Equating the corresponding energy output of the dipole to the loss rate in rotational energy, we obtain an estimate for the magnetic field strength at the pulsar surface. Typical values are of order 10^{12} G, although field strengths up to 10^{14} G have been observed (McLaughlin et al. 2003). Millisecond pulsars (MSPs) have lower field strengths of the order of 10^8 to 10^{10} G which appear to be a result of their evolutionary history. The evolution can be tracked by two parameters, the observed rotational slow-down, $\dot{P}$, and the resulting evolution in pulsar period, P. This is usually done in a (logarithmic) P–$\dot{P}$-diagram as shown in Fig. 2.

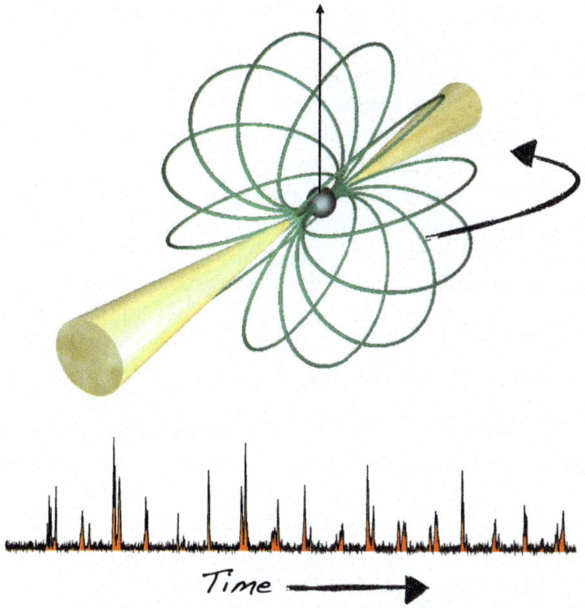

Fig. 1 Artist's impression of a rotation-powered pulsar. The magnetized neutron star appears as a pulsating source of radiation if the rotating emission beam crosses the observer's line of sight. Averaging these periodic pulses of intensity over many rotation cycles results in a stable pulse profile. Because of the timing stability of most pulsars, the arrival time of pulses can be predicted with very high precision, which is an essential requirement for all applications based on pulsar timing. (Figure by M. Kramer)

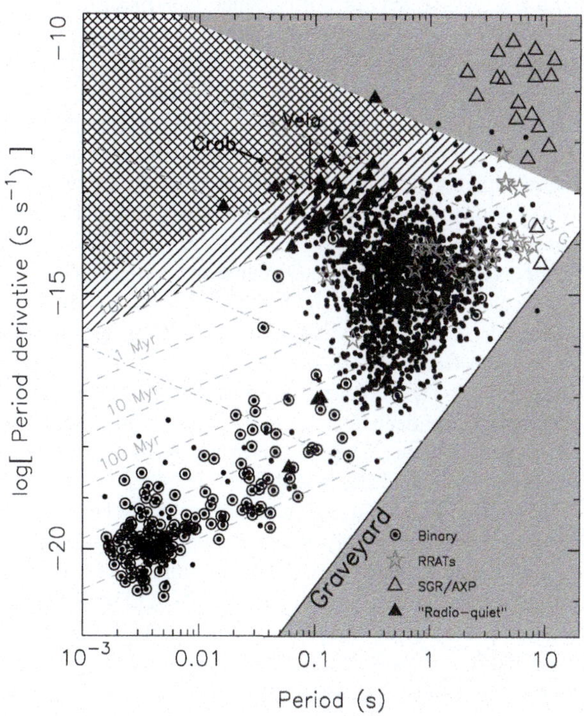

Fig. 2 The P–$\dot{P}$-diagram for the known pulsar population. Characteristic age ($\tau = P/2\dot{P}$), surface magnetic field ($B = 3.2 \times 10^{19}\sqrt{P\dot{P}}$ G) and spin-down luminosity ($\dot{E} = 4\pi I \dot{P}/P^3$, with I being the moment of inertia) are functions of P and $\dot{P}$ and are hence indicated as lines of corresponding value. Binary pulsars are marked by a *circle*. The *lower solid line* represents the pulsar "death line" enclosing the "pulsar graveyard" where pulsars are expected to switch off radio emission. The *gray area in the top right corner* indicates the region where the surface magnetic field appears to exceed the quantum critical field of 4.4×10^{13} Gauss. For such values, some theories expect the quenching of radio emission in order to explain the radio-quiet "magnetars" (i.e. Soft-gamma ray repeaters, SGRs, and Anomalous X-ray pulsars, AXPs).

Most known pulsars have spin periods between 0.1 and 1.0 s with period derivatives of typically $\dot{P} = 10^{-15}$ s s^{-1}. Selection effects are only partly responsible for the limited number of pulsars known with very long periods, the longest known period being 8.5 s

(Young et al. 1999). The dominant effect is due to the "death" of pulsars when their slow-down has reached a critical state. This state seems to depend on a combination of P and $\dot{P}$ known as the *pulsar death-line*. The normal life of radio pulsars is limited to a few tens or hundreds million years or so.

The described evolution does not explain the over 200 pulsars in the lower left of the P–$\dot{P}$-diagram (Fig. 2). These pulsars have simultaneously small periods (few milliseconds) and small period derivatives ($\dot{P} \leq 10^{-18}$ s s^{-1}). Based on their evolutionary history and characteristic ages, these *millisecond pulsars* (MSPs) are believed to be much older than ordinary pulsars with ages up to $\sim 10^{10}$ yr. MSPs evolve from pulsars with a binary companion. Once the binary companion evolves and overflows its Roche lobe, it transfers mass and thereby angular momentum (e.g. Alpar et al. 1982). In this process, previously "dead" pulsars are recycled to MSPs via an accreting X-ray binary phase. This has a number of observational consequences: (a) most normal pulsars do not develop into a MSP as they have long lost a possible companion during their violent birth event; (b) for surviving binary systems, X-ray binary pulsars represent the progenitor systems for MSPs; (c) the final spin period of recycled pulsars depends on the mass of the initial binary companion. A more massive companion evolves faster, limiting the duration of the accretion process; (d) the majority of MSPs have low-mass white-dwarf companions as the remnant of the binary star. These systems evolve from low-mass X-ray binary systems; (e) high-mass X-ray binary systems represent the progenitors for double neutron star systems (DNSs). DNSs are rare since these systems need to survive a second supernova explosion. The resulting MSP is only mildly recycled with a period of tens of millisecond. This picture explains the observation that $\sim 80\%$ of all MSPs are in a binary orbit while this is true for only less than 1% of the non-recycled population. For MSPs with a low-mass white dwarf companion the orbit is nearly circular. In case of DNSs, the orbit is affected by the unpredictable nature of the kick imparted onto the newly born neutron star in the asymmetric supernova explosion of the companion. If the system survives, the result is typically an eccentric orbit with an orbital period of a few hours. However, there is also evidence for the existence of low-kick supernova, producing DNSs with low eccentricity and relatively low-mass neutron star companions (e.g. Tauris et al. 2015).

Source with the largest estimated magnetic fields ($\sim 10^{15}$ G), the so called *magnetars*, are located in the upper right corner of Fig. 2. Here, the observed luminosity appears to exceed the neutron stars' spin-down energy loss, suggesting that magnetars in addition to the spin-down energy are powered by converting magnetic field energy (see e.g. Kaspi and Kramer 2016 for a comparison of magnetars to rotation powered radio pulsars). Only four magnetars have been detected as (transient) radio sources while all appear to be X-ray and gamma-ray sources. The long-term timing is not regular, so that applications as discussed below, are unlikely to be possible.

From timing measurements of binary MSPs (see Sect. 1.2), we can measure neutron star masses. Those are found in a range between about 1.2 and $2M_\odot$ (Özel and Freire 2016; Antoniadis et al. 2016) with the maximum mass (Antoniadis et al. 2013) ruling out the softest equation-of-sate (Demorest et al. 2010). The MSP mass distribution is strongly asymmetric. The diversity in spin and orbital properties of high-mass NSs suggests that this is most likely not a result of the recycling process, but rather reflects differences in the NS birth masses. The asymmetry is best accounted for by a bimodal distribution with a low mass component centered at $1.39M_\odot$ and a high-mass component with a mean of $1.81M_\odot$ (Antoniadis et al. 2016). These equations-of-state yield radii not too different from the very first calculations by Oppenheimer and Volkoff (1939), i.e. about 20 km in diameter, and are consistent with the blackbody emission radii determined from X-ray observations (Özel and Freire 2016). This makes neutron stars the most compact objects in the observable Universe.

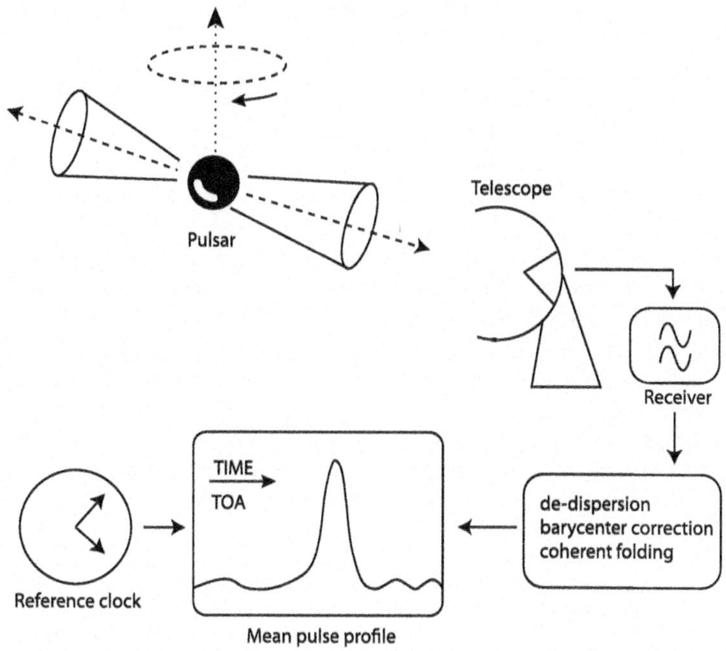

Fig. 3 Typical pulsar detection chain. The pulsar beams sweep across the radio antenna. Radio signals are recorded and analyzed in order to produce a mean pulse profile. The data processing comprises a removal of dispersion effects caused by the interstellar medium ("de-dispersion"), correction for the position and proper motion of the observatory ("barycenter correction") and coherent folding of many pulses. The time of arrival (TOA) of the pulse peak is measured against a reference clock. (Figure adapted from D. Lorimer)

1.2 Pulsar Timing: Pulsars as Clocks

The basic principle of pulsar timing is the measurement of a "time-of-arrival" (TOA) of pulses and their identification with a specific rotation number of the neutron star (cf. Fig. 3). The aim is to obtain a "coherent timing solution", where the term "coherent" refers to a complete description of the rotational phase. The experiment is repeated many times, and the measured TOAs are compared to the prediction of the timing solution. Deviations measured as timing residuals are minimized by an adjustment of the timing parameters. Eventually, the uncertainties in the rotational model are so small, that not only no rotation is missed or mislabeled between observations, but such that the timing model is able to predict the arrival time to microsecond precision or better for observations decades into the future. This is at least the case in the absence of "timing noise" (see below) and hence for the vast majority of millisecond pulsars.

Radio pulsars are usually too weak to be detected in their single pulses, so that first an average pulse must be formed to increase the signal-to-noise ratio. Single pulses also usually differ in shape, intensity and exact pulse phase (see Fig. 1), but occur within a well-defined window given by the average pulse shape. Therefore, using an average pulse also improves the timing precision and allows the usage of a technique known as template matching. With the average pulse shape expected to be constant between observations, one can compare the measured pulse form with a high signal-to-noise "template" obtained from the addition of many earlier observations by using a cross-correlation. Assuming that the discretely sampled profile, $\mathscr{P}(t)$, is a scaled and shifted version of the template, $\mathscr{T}(t)$, with added noise, $\mathscr{N}(t)$, we may write

$$\mathscr{P}(t) = a + b\mathscr{T}(t - \tau) + \mathscr{N}(t) \tag{1}$$

where a is an arbitrary offset and b a scaling factor. The time shift between the profile and the template, τ, yields the TOA relative to a fiducial point of the template and the start time of the observation (Taylor 1992). Hence, the TOA is thereby defined as the arrival time of the nearest pulse to the mid-point of the observation. The uncertainty of a TOA measurement, σ_{TOA} improves with signal-to-noise ratio (hence, size of the used telescope) and sharpness of the pulse features, as this enables a more precise cross-correlation result. For MSPs, a few thousand pulses can be added easily in a few minutes of observing time. This usually results in extremely stable profiles. In addition to their higher rotational stability and short duration pulses, this represents an important factor in explaining the superior timing stability of MSPs when compared to normal pulsars.

By measuring the arrival time of the pulsar signals very precisely, we can study effects that determine the propagation of the pulses in four-dimensional space-time. As indicated, the aim is to determine the rotation number of an observed pulse, counting from some reference epoch, t_0. We can write

$$N = N_0 + \nu_0 \times (t - t_0) + \frac{1}{2}\dot{\nu}_0 \times (t - t_0)^2 + \cdots, \tag{2}$$

where N_0 is the pulse number and ν_0 the spin frequency at the reference time, respectively. Whilst for most MSPs a second derivative, $\ddot{\nu}$, is usually too small to be measured, we expect ν and $\dot{\nu}$ to be related via the physics of the braking process,

$$\dot{\nu} = -\text{const.} \times \nu^n. \tag{3}$$

For magnetic dipole braking the *braking index* takes the value $n = 3$. If ν and its derivatives are accurately known and if t_0 coincides with the arrival of a pulse, all following pulses should appear at integer values of N — when observed in an inertial reference frame. However, our observing frame is not inertial, as we are using telescopes that are located on a rotating Earth orbiting the Sun. Therefore, we need to transfer the pulse times-of-arrival (TOAs) measured with the observatory clock (*topocentric arrival times*) to the center of mass of the solar system as the best approximation to an inertial frame available. The transformation of a topocentric TOA to such *barycentric arrival times*, t_{SSB}, is given by

$$t_{\mathrm{SSB}} = t_{\mathrm{topo}} - t_0 + t_{\mathrm{corr}} - k \times \mathrm{DM}/f_{\mathrm{obs}}^2, \tag{4}$$

$$+ \Delta_{\mathrm{Roemer},\odot} + \Delta_{\mathrm{Shapiro},\odot} + \Delta_{\mathrm{Einstein},\odot}, \tag{5}$$

$$+ \Delta_{\mathrm{Roemer,Bin}} + \Delta_{\mathrm{Shapiro,Bin}} + \Delta_{\mathrm{Einstein,Bin}}. \tag{6}$$

where DM is the so called dispersion measure, representing the integrated path length of free electron along the line of sight, and f_{obs} is the observing radio frequency (see below). We have split the transformation into three lines. The first two lines apply to every pulsar whilst the third line is only applicable to binary pulsars.

1.2.1 Clock and Frequency Corrections

The observatory time is typically maintained by local Hydrogen-maser clocks monitored by GPS signals. In a process involving a number of steps, clock corrections, t_{corr}, are retroactively applied to the arrival times in order to transfer them to a uniform atomic time ("Terrestrial Time", or TT) that would be kept by an ideal atomic clock on the geoid. It is published retroactively by the *Bureau International des Poids et Mesures* (BIPM).

The free electrons in the interstellar medium interact with the propagation radio signal, causing a frequency dependent group velocity. Consequently, the frequency components of the broadband pulse signal emitted a higher radio frequency f_{obs} arrive earlier than the corresponding low frequency components. Hence, due to this dispersion, the measured arrival time depends on the observing frequency, f_{obs} and the dispersion measure (DM). The TOA is therefore corrected for a pulse arrival at an infinitely high frequency (last term in Eq. (4)). For the best pulsars, the limiting factor in timing precision is often "interstellar weather" that causes small changes in DM and hence time-varying drifts in the TOAs. In those cases, the above term is, for instance, extended to include time-derivatives of DM which can be measured using multi-frequency observations.

1.2.2 Barycentric Corrections

The *Roemer delay*, $\Delta_{\mathrm{Roemer},\odot}$, is the classical light-travel time between the phase center of the telescope and the solar system barycenter (SSB). Given a unit vector, $\hat{s}$, pointing from the SSB to the position of the pulsar and the vector connecting the SSB to the observatory, $\vec{r}$, we find:

$$\Delta_{\mathrm{Roemer},\odot} = -\frac{1}{c}\,\vec{r}\cdot\hat{s} = -\frac{1}{c}(\vec{r}_{\mathrm{SSB}} + \vec{r}_{\mathrm{EO}})\cdot\hat{s}. \tag{7}$$

Here c is the speed of light and we have split $\vec{r}$ into two parts. The vector $\vec{r}_{\mathrm{SSB}}$ points from the SSB to the center of the Earth (geocenter). Computation of this vector requires accurate knowledge of the locations of all major bodies in the solar system and uses solar system ephemerides. The second vector $\vec{r}_{\mathrm{EO}}$, connects the geocenter with the phase center of the telescope. In order to compute this vector accurately, the non-uniform rotation of the Earth has to be taken into account, so that the correct relative position of the observatory is derived.

The *Shapiro delay*, $\Delta_{\mathrm{Shapiro},\odot}$, is a relativistic correction that corrects for an extra delay due to the curvature of space-time in the solar system (Shapiro 1964). It is largest for a signal passing the Sun's limb ($\sim 120\ \mu\mathrm{s}$) while Jupiter can contribute as much as 200 ns. In principle one has to sum over all bodies in the solar system, but in practice only the Sun is usually taken into account.

The last term in Eq. (5), $\Delta_{\mathrm{Einstein},\odot}$, is called *Einstein delay* and it describes the combined effect of gravitational redshift and time dilation due to motion of the Earth and other bodies, taking into account the variation of an atomic clock on Earth in the varying gravitational potential as it follows its elliptical orbit around the Sun (Backer and Hellings 1986).

1.2.3 Relative Motion & Shklovskii Effect

If the pulsar is moving relative to the SSB, the transverse component of the velocity, v_t, can be measured as the vector $\hat{s}$ in Eq. (7) changes with time. Present day timing precision is not sufficient to measure a radial motion although it is theoretically possible. This leaves Doppler corrections to observed periods, masses etc. undetermined. The situation changes if the pulsar has an optically detectable companion such as a white dwarf for which Doppler shifts can be measured from optical spectra.

Another effect arising from a transverse motion is the *Shklovskii effect*, also known in classical astronomy as *secular acceleration*. With the pulsar motion, the projected distance of the pulsar to the SSB is increasing, leading to an increase in any observed change of periodicity, such as pulsar spin-down or orbital decay. The observed pulse period derivative, for instance, is increased over the intrinsic value by

$$\left(\frac{\dot{P}}{P}\right)_{\text{obs}} = \left(\frac{\dot{P}}{P}\right)_{\text{int}} + \frac{1}{c}\frac{v_t^2}{d}. \tag{8}$$

For MSPs where $\dot{P}_{\text{int}}$ is small, a significant fraction of the observed change in period can be due to the Shklovskii effect.

1.2.4 Binary Pulsars

Equation (4)–(5) is used to transfer the measured TOAs to the SSB. If the pulsar has a binary companion, the light-travel time across the orbit and further relativistic effects need to be taken into account (see Eq. (6)). That adds additional orbital parameters to the set of timing parameters which have to be solved in the timing process (see below). In the simplest case, five Keplerian parameters need to be determined, i.e. orbital period, P_b; the projected semi-major axis of the orbit, $x \equiv a \sin i$ where i is the (usually unknown) orbital inclination angle; the orbital eccentricity, e; the longitude of periastron, ω; and the time of periastron passage, T_0. For a number of binary systems this Keplerian description of the orbit is not sufficient and corrections need to be applied. These can be either time derivatives of Keplerian parameters or parameters describing completely new effects (e.g. those of a Shapiro delay due to curved spacetime near the companion). In any case, it is important to note that we do not have to assume a particular theory of gravity when measuring such relativistic corrections, called "post-Keplerian" (PK) parameters (Damour and Deruelle 1985; Damour and Deruelle 1986). Instead, we can take the observational values and compare them with predictions made within a framework of specific theories of gravity (Damour and Taylor 1992).

In GR, the five most important PK parameters are given by the expressions below (Robertson 1938; Blandford and Teukolsky 1976; Damour and Deruelle 1986; Peters 1964).

$$\dot{\omega} = 3T_\odot^{2/3}\left(\frac{P_b}{2\pi}\right)^{-5/3}\frac{1}{1-e^2}(M_p+M_c)^{2/3}, \tag{9}$$

$$\gamma = T_\odot^{2/3}\left(\frac{P_b}{2\pi}\right)^{1/3} e\frac{M_c(M_p+2M_c)}{(M_p+M_c)^{4/3}}, \tag{10}$$

$$r = T_\odot M_c, \tag{11}$$

$$s = \sin i = T_\odot^{-1/3}\left(\frac{P_b}{2\pi}\right)^{-2/3} x\,\frac{(M_p+M_c)^{2/3}}{M_c}, \tag{12}$$

$$\dot{P}_b = -\frac{192\pi}{5}T_\odot^{5/3}\left(\frac{P_b}{2\pi}\right)^{-5/3} f(e)\frac{M_pM_c}{(M_p+M_c)^{1/3}}, \tag{13}$$

where all masses are expressed in solar units, G is Newton's gravitational constant, c the speed of light and

$$f(e) = \frac{(1+(73/24)e^2+(37/96)e^4)}{(1-e^2)^{7/2}}. \tag{14}$$

P_b is the period and e the eccentricity of the binary orbit. The masses M_p and M_c of pulsar and companion, respectively, are expressed in solar masses ($M_\odot$) where we define the constant $T_\odot = GM_\odot/c^3 = 4.925490947$ μs. G denotes the Newtonian constant of gravity and c the speed of light.

The first PK parameter, $\dot{\omega}$, describes the relativistic advance of periastron in rad s^{-1}. It is the easiest to measure for orbits with non-zero eccentricities (note that ω is only poorly defined for $e \approx 0$ and so is $\dot{\omega}$). From a measurement of $\dot{\omega}$, we obtain from Eq. (9) the total mass of the system, $(M_p + M_c)$.

The orbital decay due to gravitational-wave damping is expressed by the (dimensionless) change in orbital period, $\dot{P}_b$. Any metric theory of gravity that embodies Lorentz-invariance in its field equations predicts gravitational radiation and, hence, $\dot{P}_b$. If a theory satisfies the strong equivalence principle, like GR, gravitational *dipole* radiation is not expected, but *quadrupole* emission will be the lowest multipole term. In alternative theories, while the *inertial* dipole moment may remain uniform, the *gravitational wave* dipole moment may not, and dipole radiation may be predicted. The magnitude of this effect depends on the difference in gravitational binding energies, expressed by the difference in coupling constants to a scalar gravitational field.

The other two parameters, r and s, are related to the Shapiro delay caused by the curvature of space-time due to the gravitational field of the companion. They are measurable, depending on timing precision, if the orbit is seen nearly edge-on.

1.2.5 Obtaining a Timing Solution

Equations (2) to (6) contain the set of timing parameters that need to be determined. Many of them are not known a priori (or only with limited precision after the discovery of a pulsar) and need to be determined precisely in a least squares fit analysis of the measured TOAs. The parameters can be categorized into three groups: (a) *astrometric parameters* (i.e. position, proper motion, parallax contained in the Römer and Shapiro delay, respectively); (b) *spin parameters* (i.e. rotation frequency, ν, and higher derivatives); (c) *binary parameters*.

Given a minimal set of starting parameters, one uses maximum likelihood methods to match TOAs to integer numbers of pulsar rotations according to Eq. (2). For instance, we can minimize the expression

$$\chi^2 = \sum_i \left(\frac{N(t_i) - n_i}{\sigma_i} \right)^2 \tag{15}$$

where n_i is the nearest integer to $N(t_i)$ and σ_i is the TOA uncertainty in units of pulse period (turns).

In order to obtain a phase-coherent solution that accounts for every single rotation of the pulsar between two observations, one starts off with a small set of TOAs that were obtained sufficiently close in time so that the accumulated uncertainties in the starting parameters do not exceed one pulse period. Gradually, the data set is expanded, maintaining coherence in phase. When successful, post-fit residuals expressed in pulse phase show a Gaussian distribution around zero with a root mean square that is comparable to the TOA uncertainties. Incorrect or incomplete timing models cause systematic structures in the post-fit residuals identifying the parameter that needs to be included or adjusted. The precision of the parameters improves with length of the data span and the frequency of observation, but also with orbital coverage in the case of binary pulsars. See, for instance, Verbiest et al. (2016) for examples and references therein for further details.

Deviations from a regular pulsar spin-down are collectively called "timing noise". The origin of timing noise is usually identified with processes in the interior of neutron stars, observable as "glitches" (i.e. sudden increases in rotational frequency), state-changes in the pulsar magnetosphere causing a change in spin-down rate, or possibly variations in the observed pulse arrival time due to interstellar weather. The first two sources of timing noise are mostly observed in young pulsars, while recycled millisecond pulsars appear as much more stable clocks. All cases may severely affect the usage of individual sources, as we discuss in somewhat more detail in Sect. 2.2.

1.3 Pulsars as Tools

Pulsar combine a number of interesting properties: they are compact objects with strong gravitational fields; they are small enough to essentially act as test masses in binary systems; the act as natural precision clocks; their interior contains the most extreme dense matter in the observable Universe; they emit coherent radio emission that is up to 100% polarized; they are not only radio sources but pulsar emission is often seen across the whole electromagnetic spectrum. There are many more properties that one could list, but just those are sufficient to make pulsars an exciting and useful tool for probing a wide range of physics or for probing the interstellar or other surrounding media. They are especially useful for testing theories of gravity. This can be done either with pulsars as part of a binary system, or also as part of a Galactic-sized detector for low-frequency gravitational waves. We will expand on the latter in more detail below, but concentrate for a moment on binary pulsars.

While, strictly speaking, binary pulsars move in the weak gravitational field of a companion, they do provide precision tests of the (quasi-stationary) strong-field regime. This becomes clear when considering that the majority of alternative theories predicts strong self-field effects which would clearly affect the pulsars' orbital motion. Hence, tracing their fall in a gravitational potential, we can search for tiny deviations from general relativity, providing us with unique precision strong-field tests of gravity. For instance, in binary systems a wide range of relativistic effects can be observed, identified and studied (e.g. Will 2014), including concepts and principles deeply embedded in theoretical frameworks. If a specific alternative theory is developed sufficiently well, one can also use radio pulsars to test the consistency of this theory. Table 1 summarizes some, where this has been possible. In particular, the radiative properties of a theory are a very powerful and sensitive probe, so that every successful theories has to pass the binary pulsar experiments.

The tests of relativistic gravity are complementary to what is achievable with detections of gravitational waves using ground-based detectors. The latter allow access to the highly-relativistic regime which is dominated by the effects of gravitational wave emission. However, these tests are currently limited in their precision. In comparison, the Double Pulsar can test the lowest order of the general relativity's quadrupole 0 about three orders more precisely than possible with the brightest LIGO event so far (Kramer 2016). Higher order terms, in contrast, can only be tested with LIGO—at least until Pulsar Timing Arrays (see below) can make a detection. Therefore, binary pulsar experiments and direct detection of gravitational waves are highly complementary.

The precision of pulsars allow also applications that may not be directly obvious. One such application is space navigation that we will describe in detail below. Key is the clock-like stability of pulsars (Taylor 1991; Matsakis et al. 1997), allowing us to make extremely precise measurements. Table 2 gives an idea about the precision of pulsar timing that we achieve today. With larger telescopes being built (e.g. the South African MeerKAT), precision and the number applications will increase. Perhaps, at some point, we will establish a complete pulsar-based timescale.

Table 1 Constraining specific (classes of) gravity theories using radio pulsars. See Wex (2014), Kaspi and Kramer (2016) for details

Theory (class)	Method	Ref.
Scalar-tensor gravity:		
Jordan–Fierz–Brans–Dicke	Limits by PSR J1738+0333 and PSR J0348+0432, comparable to best Solar system test (Cassini)	(Freire et al. 2012)
Quadratic scalar-tensor gravity	For $\beta_0 < -3$ and $\beta_0 > 0$ best limits from PSR-WD systems, in particular PSR J1738+0333 and PSR J0348+0432	(Freire et al. 2012)
Massive Brans–Dicke	For $m_\varphi \sim 10^{-16}$ eV: PSR J1141−6545	(Alsing et al. 2012)
Vector-tensor gravity:		
Einstein–Æther	Combination of pulsars (PSR J1141−6545, PSR J0348+0432, PSR J0737−3039, PSR J1738+0333)	(Yagi et al. 2014)
Hořava gravity	Combination of pulsars (see above)	(Yagi et al. 2014)
TeVeS and TeVeS-like theories:		
Bekenstein's TeVeS	Excluded using Double Pulsar	(Kramer 2016)
TeVeS-like theories	Excluded using PSR 1738+0333	(Freire et al. 2012)

Table 2 Examples of obtained relative precision using pulsar timing. Corresponding (example) references are cited in the last column

Clock properties:		
Spin frequency	$\lesssim 10^{-15}$	(Verbiest et al. 2008)
Orbital period	$\lesssim 10^{-11}$	(Kramer 2016)
Frequency stability	$\lesssim 10^{-15}$	(Hobbs et al. 2012)
Masses measurements:		
Neutron star	$\lesssim 10^{-4}$	(Weisberg et al. 2010)
White dwarf companion	$\lesssim 10^{-3}$	(Hotan et al. 2006)
Main sequence star companion	$\lesssim 10^{-2}$	(Freire et al. 2011)
Mass of Jupiter and moons	$\lesssim 10^{-6}$	(Champion et al. 2010)
Astrometry:		
Distance	$\lesssim 10^{-2}$	(Verbiest et al. 2008)
Proper motion	$\lesssim 10^{-15}$	(Verbiest et al. 2008)
Gravity tests:		
Test of general relativity	$\lesssim 10^{-4}$	(Kramer 2016)
Constancy of grav. constant, $\dot{G}/G$	$\lesssim 10^{-12}$	(Freire et al. 2012)
GW chara. strain $\Delta L/L$	$\lesssim 10^{-16}$	(Verbiest et al. 2016)

2 Autonomous Spacecraft Navigation Based on Pulsar Timing

Possible implementations of autonomous navigation systems were already discussed in the early days of space flight (Battin 1964). In principle, the orbit of a spacecraft can be determined by measuring angles between solar system bodies and astronomical objects; e.g., the angles between the Sun and two distant stars and a third angle between the Sun and a planet. However, because of the limited angular resolution of on-board star trackers and sun sensors, this method yields spacecraft positions with uncertainties that accumulate typically to several thousand kilometers. Alternatively, the navigation fix can be established by observing multiple solar system bodies: It is possible to autonomously triangulate the spacecraft position from images of asteroids taken against a background field of distant stars. This method was realized and flight-tested on NASA's Deep-Space-1 mission between October 1998 and December 2001. The Autonomous Optical Navigation (AutoNav) system on-board Deep Space 1 provided the spacecraft orbit with 1σ errors of ± 250 km and ± 0.2 m/s, respectively (Riedel et al. 2000).

An alternative and very appealing approach to autonomous spacecraft navigation is based on pulsar timing. The idea of using these celestial sources as a natural aid to navigation goes back to the 1970s when Downs (1974) investigated the idea of using pulsating radio sources for interplanetary navigation. He analyzed a method of position determination by comparing TOAs at the spacecraft with those at a reference location. Within the limitations of technology and pulsar data available at that time (a set of only 27 radio pulsars were considered), Downs showed that spacecraft position errors on the order of 1500 km could be obtained after 24 hours of signal integration. A possible improvement in precision by a factor of 10 was estimated if better (high-gain) radio antennas were available for the observations.

Chester and Butman (1981) adopted this idea and proposed to use X-ray pulsars, of which about one dozen were known at the time. They estimated that 24 hours of data collection from a small on-board X-ray detector with $0.1\ \mathrm{m}^2$ collecting area would yield a three-dimensional position accurate to about 150 km. Their analysis, though, was not based on detailed simulations or actual pulsar timing analyses; neither did it take into account the technological requirements or weight and power constraints for implementing such a navigation system.

These early studies on pulsar-based navigation estimated relatively large position and velocity errors so that this method was not considered to be an applicable alternative to the standard navigation schemes. However, pulsar astronomy has improved considerably over the last 40 years since these early proposals. Meanwhile, pulsars have been detected across the electromagnetic spectrum and their emission properties have been studied in great detail (see Sect. 1). Along with the recent advances in detector and telescope technology this motivates a general reconsideration of the idea of pulsar-based spacecraft navigation.

2.1 The Relevance of the Various Pulsar Types for Navigation

We already discussed the various types of pulsars, namely rotation-, accretion-, and magnetic-field-powered neutron stars in Sect. 1. The very complex spin behavior and often unpredictable evolution of rotation period in accretion powered pulsars manifest themselves in erratic changes between spin-up and spin-down as well as X-ray burst activities. This unsteady and non-coherent timing behavior disqualifies them as reference sources for navigation. Similar arguments for magnetars invalidates these sources also for the use in a pulsar-based spacecraft navigation system.

Concerning their application for navigation, the only pulsar class that really qualifies is that of rotation-powered ones. Here, the much higher timing stability of MSPs is of major importance for their use in a pulsar-based navigation system. Of the ~ 2500 rotation-powered pulsars known today (Fig. 2), about 150 have been detected in the X-ray band (Becker 2009), and approximately 1/3 of them are MSPs. In the past 30–40 years many of them have been regularly timed with high precision especially in radio observations. Consequently, their ephemerides (RA, DEC, P, $\dot{P}$, binary orbit parameters, TOAs and absolute pulse phase for a given epoch, pulsar proper motion etc.) are known with very high accuracy

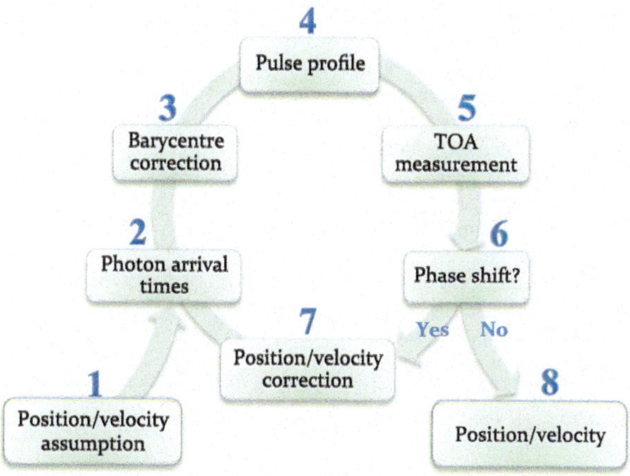

Fig. 4 Iterative determination of position and velocity by a pulsar-based navigation system (Becker et al. 2015)

(see Table 2). This is an essential requirement for using these celestial objects as navigation beacons, as it enables us to predict the TOAs of a pulsar for any location in the solar system and beyond.

2.2 Timing Irregularities

So far, we have neglected that young rotation-powered pulsars can show *glitches* in their spin-down behavior, i.e., abrupt increases of rotation frequency, often followed by an exponential relaxation toward the pre-glitch frequency (Espinoza et al. 2011; Yu et al. 2013). While this is often observed in young pulsars, it is very rarely in old and millisecond pulsars. Nevertheless, the glitch behavior of pulsars has to be taken into account by a pulsar-based navigation system. While it is possible to change the set of pulsars taken as reference when a glitch has been observed, the intrinsic *timing noise* is a factor which sets a hard limit for the overall accuracy possible for a pulsar-based spacecraft navigator. Timing noise is present in all pulsars, but the level varies between sources (Deng et al. 2012, 2013). Typical MSPs show timing residuals with an root mean square (RMS) at a level of ~ 0.1–1 μs, which limits the position accuracy then to 30–300 m.

2.3 Principles of Pulsar-Based Navigation

The concept of using pulsars as navigational aids is based on the comparison of measured TOA with predicted TOA at a given epoch and reference location. Figure 3 showed the typical chain for detecting radio signals from a rotation-powered pulsars. For our application here, we perform x-ray observations, where the dedispersion in not necessary. However, the step of applying barycenter correction to the observed photon arrival times (see Sect. 1.2.2) is still essential. The pulsar ephemerides along with the position and velocity of the observer are parameters of this correction. Using a spacecraft position that deviates from the true position during the observation results in a phase shift of the pulse peak (or equivalently in a difference in the pulse arrival time). Therefore, the position and velocity of the spacecraft can be adjusted in an iterative process until the pulse arrival time matches with the expected one. The corresponding iteration chain is shown in Fig. 4.

An initial assumption of position and velocity is given by the planned orbital parameters of the spacecraft (1). The iteration starts with a pulsar observation, during which the

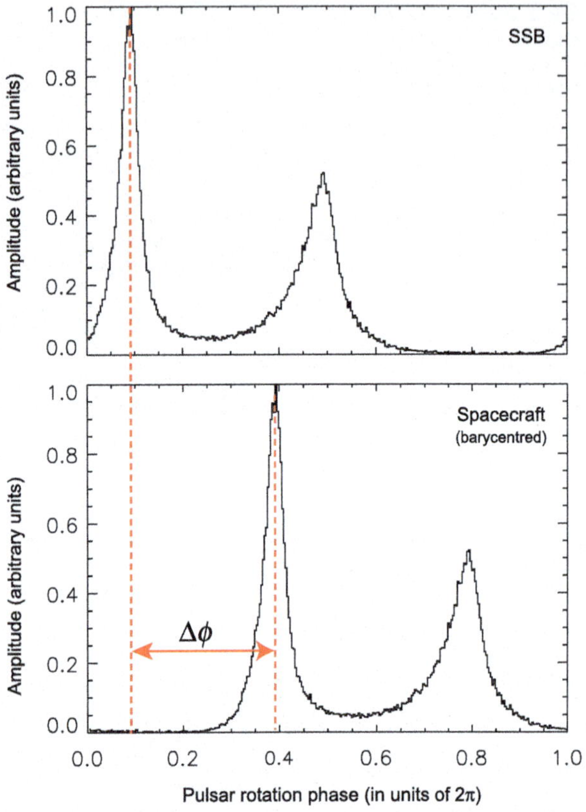

Fig. 5 Measuring the phase difference between the expected and measured pulse peak at an inertial reference location; e.g., the solar system barycenter (SSB). The *top profile* shows the main pulse peak location as expected at the SSB. The *bottom profile* is the one measured at a spacecraft and transformed to the SSB by assuming the spacecraft position and velocity during the observation. If the position and velocity assumption was wrong, a phase shift $\Delta\phi$ is observed (Becker et al. 2015)

arrival time of individual photons are recorded (2). The photon arrival times have to be corrected for the proper motion of the spacecraft by transforming the arrival times to an inertial reference location; e.g., the solar system barycenter (SSB). As for pulsar observations with terrestrial telescopes, this correction requires knowledge of the (assumed or deduced) spacecraft position and velocity as input parameters. The barycenter corrected photon arrival times allow then the construction of a pulse profile or pulse phase histogram (4) representing the temporal emission characteristics and timing signature of the pulsar. This pulse profile, which is continuously improving in significance during an observation, is permanently correlated with a pulse profile template in order to increase the accuracy of the absolute pulse-phase measurement (see Sect. 1.2) (5), or equivalently, TOA. From the pulsar ephemeris that includes the information of the absolute pulse phase for a given epoch, the phase difference $\Delta\phi$ between the measured and predicted pulse phase can be determined (cf. Fig. 5).

In this scheme, a phase shift (6) with respect to the absolute pulse phase corresponds to a range difference $\Delta x = cP(\Delta\phi + n)$ along the line of sight toward the observed pulsar. Here, $\Delta\phi$ the phase shift and $n = 0, \pm1, \pm2, \ldots$ an integer that takes into account the periodicity of the observed pulses. If the phase shift is non-zero, the position and velocity of the spacecraft needs to be corrected accordingly and the next iteration step is taken (7). If the phase shift is zero, or falls below a certain threshold, the position and velocity used during the barycenter correction was correct (8) and corresponds to the actual orbit of the spacecraft.

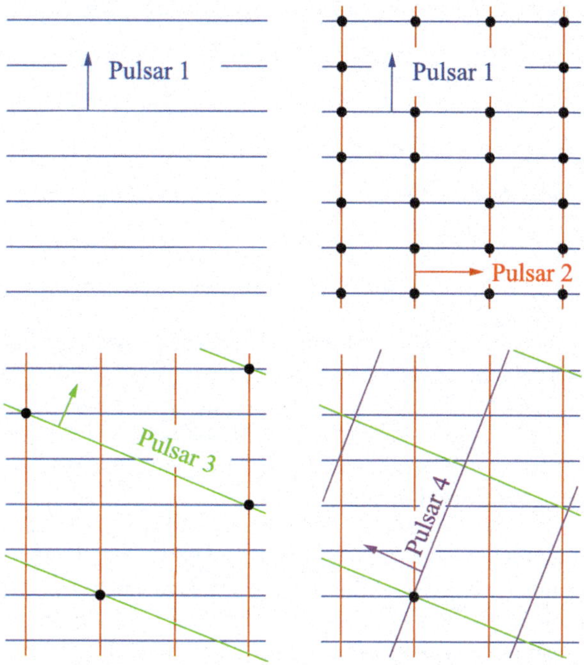

Fig. 6 Solving the ambiguity problem by observing four pulsars (drawn in two dimensions). The *arrows* point along the pulsar's lines-of-sight. *Straight lines* represent planes of constant pulse phase; *black dots* indicate intersections of planes (Becker et al. 2015)

A three-dimensional position fix can be derived from observations of at least three different pulsars (cf. Fig. 6). If on-board clock calibration is necessary, the observation of a fourth pulsar is required.

Since the position of the spacecraft is deduced from the phase (or TOA) of a periodic signal, ambiguous solutions may occur. This problem can be solved by constraining the domain of possible solutions to a finite volume around an initial assumed position (Bernhardt et al. 2010, 2011), or by observing additional pulsars as illustrated in Fig. 6.

3 Gravitational Wave Detection Based on Pulsar Timing

3.1 Principles of Gravitational Wave Detection with Pulsar Timing

Because of their exquisite stability, MSPs also constitute a formidable tool for the detection of low frequency gravitational waves (GWs). The underlying concept is very simple: GWs affect the propagation of radio signals from the pulsar to the Earth, leaving a characteristic signature in the pulses TOAs. If this effect is not included in the timing solutions described in Sect. 1.2, it will show up in pulsar "a" as a residual of the form (Sesana and Vecchio 2010):

$$r_a(t) = \int_0^t \frac{\delta\nu_a}{\nu_a}(t')dt', \tag{16}$$

where

$$\frac{\delta\nu_a}{\nu_a} = \frac{1}{2}\frac{\hat{p}_a^i\hat{p}_a^j}{1+\hat{p}_a.\hat{\Omega}}\Delta h_{ij}. \tag{17}$$

Here ν_a is the MSP spin frequency, $\hat{p}_a$ denotes the position of the pulsar on the sky, and $\hat{\Omega}$ is the direction of the GW propagation. The quantity

$$\Delta h_{ij} = h_{ij}(t_a^P) - h_{ij}(t) \tag{18}$$

is the strain of the GW at the location of the pulsar $h_{ij}(t_a^P)$ and on Earth $h_{ij}(t)$ (indices i, j run on the three spatial coordinates), and the pulsar time t_a^P is related to the Earth time t as:

$$t_a^P = t - L_a(1 + \hat{\Omega}.\hat{p}_a) \equiv t - \tau_a, \tag{19}$$

where L_a is the distance to the pulsar. In practice, GWs cause drifts in the MSP spin frequency (redshift/blueshift) proportional to the difference of the perturbing field at the pulsar and on Earth. Those drifts cause the pulse to arrive at the Earth earlier/later than expected, resulting in the distinctive timing residual $r_a(t)$ of Eq. (16). The properties of $r_a(t)$ ultimately depend on the nature of the perturbation Δh_{ij} which can be either a deterministic function or a stochastic process. Both cases are astrophysically relevant, and analysis methods have been developed accordingly.

Pulsars are complex astrophysical objects, governed by largely unknown physics that might affect their intrinsic stability and the GW signal has to be disentangled from a plethora of other effects (see Lentati et al. 2015 for a comprehensive discussion). For example, a deterministic sinusoidal GW signal might be virtually indistinguishable from the effect of a binary companion. Therefore, the GW residual signature needs to be simultaneously detected in several MSPs. This is even more crucial if the GW signal has a stochastic nature. In this case, $r_a(t)$ is just a realization of a stochastic process, at par with other stochastic noise sources. GW detection therefore requires the cross correlation of the TOAs from an ensemble of MSPs, forming a pulsar timing array (PTA, Foster and Backer 1990). Three main PTA collaborations are now active in the monitoring of several tens of MSPs: The European Pulsar Timing Array (EPTA, Desvignes et al. 2016), the Parkes Pulsar Timing Array (PPTA, Reardon et al. 2016) and the North American Nanohertz Observatory for Gravitational Waves (NANOGrav, Arzoumanian et al. 2015). The three collaborations are constantly improving their data, that they also share since 2010 under the aegis of the International Pulsar Timing Array (IPTA, Verbiest et al. 2016), aiming at the formation of a combined, more sensitive dataset.

Altogether, the three PTAs are timing order of fifty of the best MSPs at a weekly cadence (Δt) for a timespan T of several years (more then 20 for some cases), with a timing precision of few micro-seconds to few tens of nano-seconds. PTAs are therefore sensitive to GWs in the frequency range $1/T < f < 1/(2\Delta t)$, corresponding to few to few-hundred nano-Hertz. Possible GW sources in this frequency range include cosmological stochastic backgrounds from inflation, phase transitions or cosmic strings (Lasky et al. 2016), but the loudest signals are expected from the cosmic population of inspiralling supermassive black hole binaries (SMBHBs) formed following galaxy mergers (Sesana et al. 2008).

3.2 Signals from Supermassive Black Hole Binaries

In the astrophysically reasonable assumption of circular, monochromatic, non-precessing binary, orbiting at a frequency f, the two independent polarization amplitudes generated by the system can be written as (Sesana and Vecchio 2010):

$$h_+(t) = \left(1 + \cos^2 \iota\right) A_{\rm gw} \cos \Phi(t), \tag{20a}$$

$$h_{\times}(t) = -2\cos\iota\, A_{\rm gw}\sin\Phi(t)\,, \tag{20b}$$

where

$$A_{\rm gw}(f) = 2\frac{\mathcal{M}^{5/3}}{D}\left[\pi f(t)\right]^{2/3} \tag{21}$$

is the GW amplitude, D the luminosity distance to the GW source, $\mathcal{M} = (M_1 M_2)^{3/5}/(M_1 + M_2)^{1/5}$ is the chirp mass (being M_1 and M_2 the masses of the two SMBHs), $\Phi(t) = 2\pi\int^t f(t')dt'$ is the GW phase, and ι is the inclination angle (i.e. the angle defined by the direction of the orbital angular momentum of the binary and the line of sight to the observer. The metric perturbation in Eq. (18) can therefore be written as:

$$h_{ij}(t,\hat{\Omega}) = e^{+}_{ij}(\hat{\Omega})\big(h_{+}(t^{P}_{a},\hat{\Omega}) - h_{+}(t,\hat{\Omega})\big) + e^{\times}_{ij}(\hat{\Omega})\left(h_{\times}(t^{P}_{a},\hat{\Omega}) - h_{\times}(t,\hat{\Omega})\right), \tag{22}$$

where the contribution of both the pulsar and the Earth term have been included. $e^{A}_{ij}(\hat{\Omega})$ $(A = +, \times)$ are the polarization tensors, that are uniquely defined once one specifies the wave principal axes described by the unit vectors $\hat{m}$ and $\hat{n}$ as,

$$e^{+}_{ij}(\hat{\Omega}) = \hat{m}_i\hat{m}_j - \hat{n}_i\hat{n}_j, \tag{23a}$$

$$e^{\times}_{ij}(\hat{\Omega}) = \hat{m}_i\hat{n}_j + \hat{n}_i\hat{m}_j. \tag{23b}$$

The signal is therefore deterministic and can be described as $r_a(\vec{\lambda}, t)$, where $\vec{\lambda}$ is the vector of parameters specifying the SMBHB (including SMBH masses, binary sky location, inclination, frequency and initial phase). Note that r_a also depends on the sky location of the pulsar through the response function in Eq. (17) and on the distance to the pulsar that affects the phase and possibly the frequency of the pulsar term (see Babak et al. 2016 for a full description).

Since galaxy mergers are common, we expect *at any time* a large population of SMBHB emitting GWs in the PTA band. Therefore, the superposition of many incoherent signals results in a stochastic GW background (GWB), with energy content described in terms of its GW energy density $\rho_{\rm gw}$ per unit logarithmic frequency, divided by the critical energy density, ρ_c, to close the Universe:

$$\Omega_{\rm gw}(f) = \frac{1}{\rho_c}\frac{d\rho_{\rm gw}}{d\ln f} = \frac{2\pi^2}{3H_0^2}f^2 h_c^2(f). \tag{24}$$

Here, f is the GW frequency, $\rho_c = 3H_0^2/8\pi$ is the critical energy density required to close the Universe, $H_0 = 100\ h$ km s^{-1} Mpc^{-1} is the Hubble expansion rate. In the limit of circular GW-driven SMBHBs the 'characteristic strain', $h_c(f)$, associated with a GWB energy density $\Omega_{\rm gw}(f)$ is parametrized as a single power-law:

$$h_c = A\left(\frac{f}{\rm yr^{-1}}\right)^{-2/3}, \tag{25}$$

where A is the strain amplitude at a characteristic frequency of 1 yr^{-1}. Finally, h_c is directly related to the observable quantity induced by a GWB in the timing residuals, the one-sided power spectral density, $S(f)$, given by:

$$S(f) = \frac{1}{12\pi^2}\frac{1}{f^3}h_c(f)^2 = \frac{A^2}{12\pi^2}\left(\frac{f}{\rm yr^{-1}}\right)^{-13/3} {\rm yr}^3. \tag{26}$$

Therefore, in the case of a stochastic GWB, the signal is described by a stochastic red process with power spectral density given by Eq. (26). This can be described by a vector of two parameters $\vec{\eta} = (A, \gamma)$, where $\gamma = -13/3$ for circular GW-driven SMBHBs. We will see below that this power has a very specific correlations between pairs of pulsars, which is the distinctive signature that has to be searched in PTA data.

3.3 Current Status: Upper Limits from Pulsar Timing Arrays

Whether the signal is deterministic or stochastic, the core aspect of all GW searches is the evaluation of the likelihood that some signal is present in the time series of the pulsar TOAs. Without entering in technical details (see, e.g. van Haasteren and Levin 2013; Lentati et al. 2015) the likelihood marginalized over the timing parameters can be written as

$$\mathcal{L}(\vec{\theta}, \vec{\lambda}, \vec{\eta} \,|\, \vec{\delta t}) = \frac{1}{\sqrt{(2\pi)^{n-m} det(G^T C G)}} \times \exp\left(-\frac{1}{2}(\vec{\delta t} - \vec{r}(\vec{\lambda}))^T G (G^T C(\vec{\eta}) G)^{-1} G^T (\vec{\delta t} - \vec{r}(\vec{\lambda}))\right). \tag{27}$$

Here n is the length of the vector $\vec{\delta t} = \bigcup \delta t_a$ obtained by concatenating the individual pulsar TOA series δt_a, m is the total number of parameters describing the timing model (see Sect. 1.2), and the matrix G is related to the design matrix (see van Haasteren and Levin 2013 for details). The variance-covariance matrix C, in its more general version, contains contributions from the GWB and from the white and (in general) red noise: $C = C_{gw}(\vec{\eta}) + C_{wn} + C_{rn}$. We refer the reader to Lentati et al. (2015) for exact expressions of the noise variance-covariance matrix.

Despite of its complexity, Eq. (27) is conceptually not distant from the more familiar Eq. (15). In obtaining a timing solution one describes the 'distance' between the data and the model in terms of χ^2, and adopts as best solution the set of parameters minimizing this distance. The inverse of the χ^2 can be thought as a likelihood function, and the solution of the problem is the maximum likelihood point. Likewise, the multivariate Gaussian likelihood in Eq. (27) is a measure of how well the data is described by some parametric model for resolvable sources ($\vec{\lambda}$ parameters) and a stochastic GWB ($\vec{\eta}$ parameters). Conversely, The timing model parameters—those obtained in Eq. (15)—are marginalized over using the design matrix G. One can then find the set of model parameters that maximizes the likelihood, or compare the likelihoods for models with and without a specific GW source to decide whether the data favor the presence of a signal, etc. Note that, since the fingerprint of a stochastic GWB is hidden in the specific correlation pattern found in the residuals collected from an ensemble of sources, it is necessary to describe the correlation properties of the residual via the computation of the variance-covariance matrix C. Although this implies a significant computational burden, it does not change the nature of the problem.

Likelihoods similar to Eq. (27) have been used to search for both (i) a deterministic single GW sources (e.g., particularly loud SMBHBs rising above the level of the stochastic GWB), or (ii) a stochastic GWB.

In case (i), a deterministic signal $r(\vec{\lambda}, t)$ is added to the model and a search is performed over the parameter space defined by $\vec{\lambda}$. If the data are better described by noise plus a deterministic signal, then the properties of the latter can be inferred through the posterior distribution of $\vec{\lambda}$; otherwise, upper limits on the amplitude A_{gw} of a single GW signal

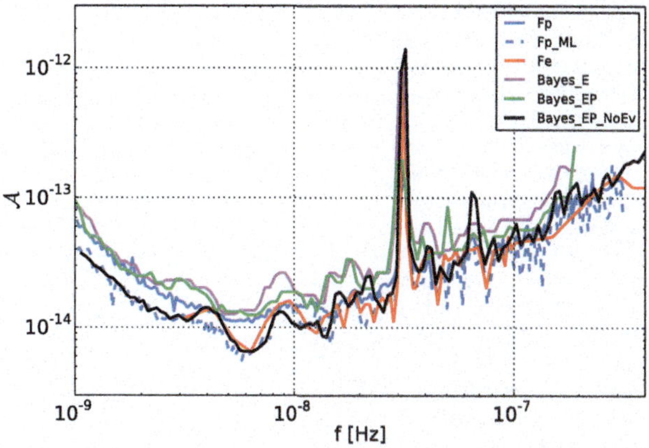

Fig. 7 The 95% upper limit on the gravitational wave strain A_{gw} for the 3 frequentist methods, i.e. $\mathscr{F}_p$ varying noise (*Fp*), $\mathscr{F}_p$ fixed noise (*Fp_ML*) and $\mathscr{F}_e$, and the 3 Bayesian methods, i.e. "evolving source" with Earth term only (*Bayes_E*) and with Earth and Pulsar terms (*Bayes_EP*) and "non-evolving source" with Earth and Pulsar terms (*Bayes_EP_NoEv*), performed in the EPTA single source analysis. From Babak et al. (2016), where the detailed descriptions of each method can be found

(Eq. (21)) can be placed at every frequency. Searches of this type have been performed by the three major PTAs (Arzoumanian et al. 2014; Zhu et al. 2014; Babak et al. 2016). The most stringent limit to date has been placed by EPTA and is shown in Fig. 7 (from Babak et al. 2016) as a function of frequency for different type of searches (see Babak et al. 2016 for full details). Searches yielded yet no convincing evidence of individual GW sources, and the presence of SMBHBs of few billion solar masses emitting in the PTA range can be ruled out to the distance of the Coma cluster.

In case (ii), i.e. the search for a GWB, $C_{gw}(\vec{\eta})$ is considered in the likelihood evaluation. The signal is stochastic in nature, and has to be singled out among other stochastic sources of noise, including red noise processes peculiar to individual MSPs, correlated noises due to clock or ephemeris errors etc. The smoking gun of a stochastic GWB is provided by the peculiar correlation pattern it introduces in the residuals of pulsar pairs as a function of their separation. This is a general property of any GWB described by the two tensor polarizations allowed by GR and was first worked out by Hellings and Downs (1983):

$$\Gamma(\zeta_{IJ}) = \frac{3}{8}\left[1 + \frac{\cos\zeta_{IJ}}{3} + 4(1-\cos\zeta_{IJ})\ln\left(\sin\frac{\zeta_{IJ}}{2}\right)\right](1+\delta_{IJ}). \tag{28}$$

Here ζ_{IJ} is the angle between the pulsars I and J on the sky and $\Gamma(\zeta_{IJ})$ is the overlap reduction function, which represents the expected correlation between the TOAs given an isotropic stochastic GWB, and the δ_{IJ} term accounts for the pulsar term for the autocorrelation.

In the Fourier representation of the covariance matrix $C_{\rm GW}$, this results in a contribution (Lentati et al. 2015)

$$\Psi_{I,J,i,j} = \Gamma(\zeta_{IJ})\varphi_i^{\rm GWB}\delta_{ij}, \tag{29}$$

that can be included in the likelihood function (27), appropriately evaluated in the Fourier domain. Lower case indices i, j run over the difference frequency bins of the Fourier decomposition and φ_i corresponds to the power in the GW signal given by Eq. (26), evaluated at the central frequency f_i if the i-th bin. Also in this case, one can evaluate whether the data are better described by including a GWB (and therefore evaluate the vector parameter $\vec{\eta}$) or not, and in the latter case upper limits on $S_h(f)$ (or equivalently on $h_c(f)$) can be placed as a factor of frequency. Also in this case, no convincing detection has been made yet. An illustrative example is given in Fig. 8, from Lentati et al. (2015), where the

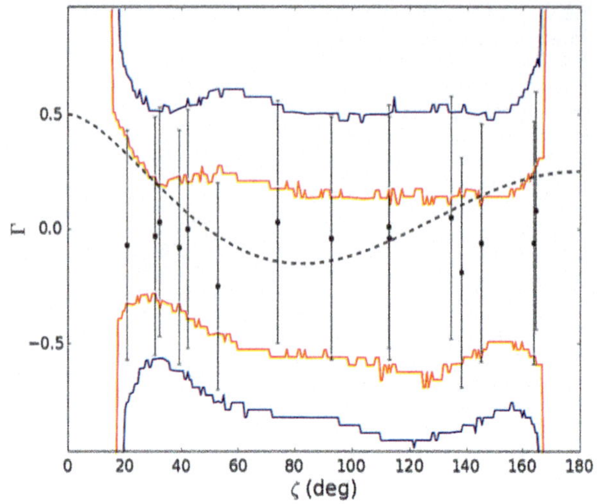

Fig. 8 The recovered correlation between pulsars as a function of angular separation on the sky for a power law noise process in the EPTA analysis. The *red* and *blue lines* represent the 68% and 95% confidence intervals for the correlation function when modeled by the lowest 4 Chebyshev polynomials, while the individual points are the mean correlation coefficient with 1σ uncertainty for each pulsar pair when fitting without assuming a smooth model. The Hellings–Downs correlation is represented by the *dotted line*. From Lentati et al. (2015)

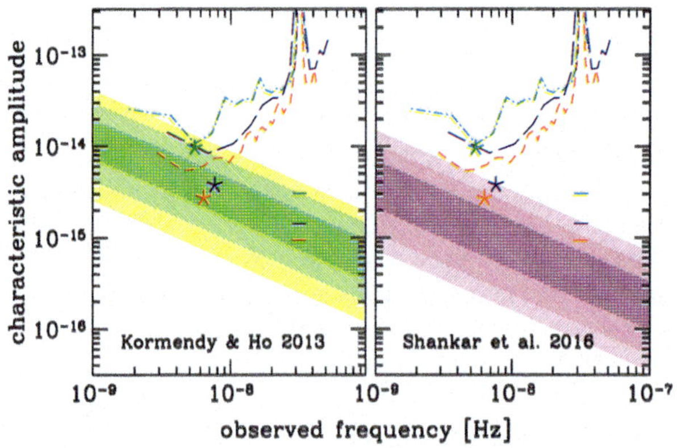

Fig. 9 From Sesana et al. 2016. PTA limits on the amplitude of a stochastic GWB. The jagged curves are current PTA sensitivities: EPTA (*dot-dashed green*), NANOGrav (*long-dashed blue*), and PPTA (*short-dashed red*). For each sensitivity curve, *stars* represent the integrated upper limits to an $f^{-2/3}$ background, and the *horizontal ticks* are their extrapolation at $f = 1\ \mathrm{yr}^{-1}$, i.e. the upper limit on $A(f = 1\ \mathrm{yr}^{-1})$ quoted in the main text. The *shaded areas* represent the 68% 95% and 99.7% confidence intervals of the characteristic amplitude $h_c(f)$ of the expected GWB form SMBHBs. The *two panels* represent predictions from models employing two different M_{BH}-host bulge mass relations: the one from Kormendy and Ho 2013 (*left*) and the one from Shankar et al. 2016 (*right*). See Sesana et al. (2016) for full details of the employed models

measured correlation pattern in the EPTA dataset is shown as a function of the angular separation of the MSP pairs used in the analysis. As expected, the correlation is consistent with zero and the functional form (28) could not be detected. With a null result, limits on $h_c(f)$ can be placed, as shown in Fig. 9 from Sesana et al. (2016). Upper limits are usually quoted in term of the amplitude A at $f = \mathrm{yr}^{-1}$, as defined in Eq. (26). Current limits are $A = 3 \times 10^{-15}$, 1.5×10^{-15}, 10^{-15}, 1.7×10^{-15} for EPTA (Lentati et al. 2015), NANOGrav (Arzoumanian et al. 2016), PPTA (Shannon et al. 2015) and IPTA (Verbiest et al. 2016) respectively. Note that, although the IPTA limit is not the most stringent one, it has not been obtained combining the most recent individual PTA datasets. Compared to the individual dataset used in the combination, the IPTA limit is a factor of ≈ 2 better, demonstrating the great potential of collaborating at a worldwide level. As shown in Fig. 9, these limits start to dig into the level expected by a cosmological population of SMBHBs, and a positive detection might be expected in the near future.

4 Summary and Future Prospects

Pulsars are unique astrophysical sources that combine a number of special properties, which converts them into effective cosmic laboratories for fundamental physics. After having provided the very first evidence for GWs (Taylor 1994), pulsars can also be used as GW detectors, probing a nHz-frequency range, where we expect the signals from supermassive binary black holes to occur.

In fact, following galaxy mergers, the SMBHs hosted at their centers pair together forming a binary system. Since galaxy mergers are observed to be frequent in the Universe, a large population of adiabatically inspiralling SMBHBs is expected to fill the nHz GW sky. Such low frequencies are inaccessible to ground and space based GW interferometers, but are in the reach of ongoing and forthcoming PTAs.

Current PTAs are already placing stringent limits to the maximum GW strain emitted by these systems, starting to pierce through the level predicted by current SMBH assembly models. As timing continues, PTA sensitivity naturally increases and progressively extends at lower frequencies (where the signal is expected to be the loudest). Moreover, new instrumentation will play a pivotal role in the coming years. MeerKAT in South Africa and FAST in China will provide even better timing precision on a larger ensemble of pulsars, and by the mid 20s' we will be surveying thousands of MSPs throughout the galaxy with SKA.

With these premises, it is likely that nHz GW detection will be likely within the next decade, placing another milestone in our understanding of the cosmic evolution of the biggest compact objects in the Universe.

But the clock-like properties of pulsars, and their accessibility across the whole electromagnetic spectrum, also open the door for a wide rage of practical applications. Here we have especially discussed spacecraft navigation.

The knowledge of how to use stars, planets and stellar constellations for navigation was fundamental for mankind in discovering new continents and subduing living space in ancient times. It is fascinating to see how history repeats itself in that a special population of stars may play again a fundamental role in the future of mankind by providing a reference for navigating their spaceships through the Universe.

Autonomous spacecraft navigating with pulsars is feasible when using either phased-array radio antennas of at least $150\,\mathrm{m}^2$ antenna area or compact light-weighted X-ray telescopes and detectors, which are currently developed for the next generation of X-ray observatories.

Using the X-ray signals from MSPs, Becker et al. (2015) estimated that navigation would be possible with an accuracy of ± 5 km in the solar system and beyond. The uncertainty is dominated by the inaccuracy of the X-ray pulse profile templates that were used for the pulse peak fittings and pulse-TOA measurements. As those are known with much higher accuracy in the radio band, it is possible to increase the accuracy of pulsar navigation down to the meter scale by using radio signals from pulsars for navigation.

The disadvantage of radio observations in a navigation application, though, is the large size and mass of the phased-antenna array. The antenna area is inversely proportional to the square root of the integration time; i.e., the same signal quality can be obtained with a reduced antenna size by increasing the observation time. However, the observing time is limited by the Allen variance of the receiving system and, therefore, cannot become arbitrarily large. In addition, irradiation from the on-board electronics requires an efficient electromagnetic shielding to prevent signal feedback. This shielding will further increase the navigator weight in addition to the weight of the antenna.

The optimal choice of the observing band depends on the boundary conditions given by a specific mission. What power consumption and what navigator weight might be allowed for may determine the choice for a specific wave band.

In general, however, it is clear already today that this navigation technique will find its applications in future astronautics. The technique behind it is very simple and straightforward, and pulsars are available everywhere in the Galaxy. Today ≈ 2500 pulsars are known. With the next generation of radio observatories, like the SKA, it is expected to detect signals from about 20 000 to 30 000 pulsars (Smits et al. 2009).

Finally, pulsar-based navigation systems can operate autonomously. This is one of their most important advantages, and is interesting also for current space technologies; e.g., as augmentation of existing GPS/Galileo satellites. Future applications of this autonomous navigation technique might be on planetary exploration missions and on manned missions to Mars or beyond.

Acknowledgement Open access funding provided by Max Planck Society. A.S. is supported by a University Research Fellowship of the Royal Society.

References

M.A. Alpar, A.F. Cheng, M.A. Ruderman, J. Shaham, Nature **300**, 728 (1982)

J. Alsing, E. Berti, C.M. Will, H. Zaglauer, Phys. Rev. D **85**(6), 064041 (2012)

J. Antoniadis, P.C.C. Freire, N. Wex, T.M. Tauris, R.S. Lynch, M.H. van Kerkwijk, M. Kramer, C. Bassa, V.S. Dhillon, T. Driebe, J.W.T. Hessels, V.M. Kaspi, V.I. Kondratiev, N. Langer, T.R. Marsh, M.A. McLaughlin, T.T. Pennucci, S.M. Ransom, I.H. Stairs, J. van Leeuwen, J.P.W. Verbiest, D.G. Whelan, Science **340**, 448 (2013)

J. Antoniadis, T.M. Tauris, F. Ozel, E. Barr, D.J. Champion, P.C.C. Freire, ArXiv e-prints (2016)

Z. Arzoumanian, A. Brazier, S. Burke-Spolaor, S.J. Chamberlin, S. Chatterjee, J.M. Cordes, P.B. Demorest, X. Deng, T. Dolch, J.A. Ellis, R.D. Ferdman, N. Garver-Daniels, F. Jenet, G. Jones, V.M. Kaspi, M. Koop, M.T. Lam, T.J.W. Lazio, A.N. Lommen, D.R. Lorimer, J. Luo, R.S. Lynch, D.R. Madison, M.A. McLaughlin, S.T. McWilliams, D.J. Nice, N. Palliyaguru, T.T. Pennucci, S.M. Ransom, A. Sesana, X. Siemens, I.H. Stairs, D.R. Stinebring, K. Stovall, J. Swiggum, M. Vallisneri, R. van Haasteren, Y. Wang, W.W. Zhu (NANOGrav Collaboration), Astrophys. J. **794**, 141 (2014). https://doi.org/10.1088/0004-637X/794/2/141

Z. Arzoumanian, A. Brazier, S. Burke-Spolaor, S. Chamberlin, S. Chatterjee, B. Christy, J.M. Cordes, N. Cornish, K. Crowter, P.B. Demorest, T. Dolch, J.A. Ellis, R.D. Ferdman, E. Fonseca, N. Garver-Daniels, M.E. Gonzalez, F.A. Jenet, G. Jones, M.L. Jones, V.M. Kaspi, M. Koop, M.T. Lam, T.J.W. Lazio, L. Levin, A.N. Lommen, D.R. Lorimer, J. Luo, R.S. Lynch, D. Madison, M.A. McLaughlin, S.T. McWilliams, D.J. Nice, N. Palliyaguru, T.T. Pennucci, S.M. Ransom, X. Siemens, I.H. Stairs, D.R. Stinebring, K. Stovall, J.K. Swiggum, M. Vallisneri, R. van Haasteren, Y. Wang, W. Zhu (The NANOGrav Collaboration), Astrophys. J. **813**, 65 (2015). https://doi.org/10.1088/0004-637X/813/1/65

Z. Arzoumanian, A. Brazier, S. Burke-Spolaor, S.J. Chamberlin, S. Chatterjee, B. Christy, J.M. Cordes, N.J. Cornish, K. Crowter, P.B. Demorest, X. Deng, T. Dolch, J.A. Ellis, R.D. Ferdman, E. Fonseca, N. Garver-Daniels, M.E. Gonzalez, F. Jenet, G. Jones, M.L. Jones, V.M. Kaspi, M. Koop, M.T. Lam, T.J.W. Lazio, L. Levin, A.N. Lommen, D.R. Lorimer, J. Luo, R.S. Lynch, D.R. Madison, M.A. McLaughlin,

S.T. McWilliams, C.M.F. Mingarelli, D.J. Nice, N. Palliyaguru, T.T. Pennucci, S.M. Ransom, L. Sampson, S.A. Sanidas, A. Sesana, X. Siemens, J. Simon, I.H. Stairs, D.R. Stinebring, K. Stovall, J. Swiggum, S.R. Taylor, M. Vallisneri, R. van Haasteren, Y. Wang, W.W. Zhu (NANOGrav Collaboration), Astrophys. J. **821**, 13 (2016). https://doi.org/10.3847/0004-637X/821/1/13

S. Babak, A. Petiteau, A. Sesana, P. Brem, P.A. Rosado, S.R. Taylor, A. Lassus, J.W.T. Hessels, C.G. Bassa, M. Burgay, R.N. Caballero, D.J. Champion, I. Cognard, G. Desvignes, J.R. Gair, L. Guillemot, G.H. Janssen, R. Karuppusamy, M. Kramer, P. Lazarus, K.J. Lee, L. Lentati, K. Liu, C.M.F. Mingarelli, S. Osłowski, D. Perrodin, A. Possenti, M.B. Purver, S. Sanidas, R. Smits, B. Stappers, G. Theureau, C. Tiburzi, R. van Haasteren, A. Vecchio, J.P.W. Verbiest, Mon. Not. R. Astron. Soc. **455**, 1665 (2016). https://doi.org/10.1093/mnras/stv2092

D.C. Backer, R.W. Hellings, Annu. Rev. Astron. Astrophys. **24**, 537 (1986)

R.H. Battin, *Astronautical Guidance* (McGraw–Hill, New York, 1964)

W. Becker, in *Neutron Stars and Pulsars*, ed. by W. Becker. Astrophysics and Space Science Library, vol. 357 (Springer, Berlin, 2009), pp. 91–140

W. Becker, M.G. Bernhardt, A. Jessner, Astron. Nachr. **336**, 749 (2015). https://doi.org/10.1002/asna.201512251

M.G. Bernhardt, T. Prinz, W. Becker, U. Walter, in *High Time Resolution Astrophysics IV, PoS(HTRA-IV)050* (2010), pp. 1–5

M.G. Bernhardt, W. Becker, T. Prinz, F.M. Breithuth, U. Walter, in *2nd International Conference on Space Technology* (2011), pp. 1–4

R. Blandford, S.A. Teukolsky, Astrophys. J. **205**, 580 (1976)

D.J. Champion, G.B. Hobbs, R.N. Manchester, R.T. Edwards, D.C. Backer, M. Bailes, N.D.R. Bhat, S. Burke-Spolaor, W. Coles, P.B. Demorest, R.D. Ferdman, W.M. Folkner, A.W. Hotan, M. Kramer, A.N. Lommen, D.J. Nice, M.B. Purver, J.M. Sarkissian, I.H. Stairs, W. van Straten, J.P.W. Verbiest, D.R.B. Yardley, Astrophys. J. **720**, L201 (2010)

T.J. Chester, S.A. Butman, Navigation Using X-Ray Pulsars. NASA Tech. Rep. 81N27129, Jet Propulsion Laboratory, Pasadena, CA, USA (1981)

T. Damour, N. Deruelle, Ann. Inst. Henri Poincaré A, Phys. Théor. **43**, 107 (1985)

T. Damour, N. Deruelle, Ann. Inst. Henri Poincaré A, Phys. Théor. **44**, 263 (1986)

T. Damour, J.H. Taylor, Phys. Rev. D **45**, 1840 (1992)

P.B. Demorest, T. Pennucci, S.M. Ransom, M.S.E. Roberts, J.W.T. Hessels, Nature **467**, 1081 (2010)

X.P. Deng, W. Coles, G. Hobbs et al., Mon. Not. R. Astron. Soc. **424**, 244 (2012)

X.P. Deng, G. Hobbs, X.P. You, Adv. Space Res. **52**, 1602 (2013)

G. Desvignes, R.N. Caballero, L. Lentati, J.P.W. Verbiest, D.J. Champion, B.W. Stappers, G.H. Janssen, P. Lazarus, S. Osłowski, S. Babak, C.G. Bassa, P. Brem, M. Burgay, I. Cognard, J.R. Gair, E. Graikou, L. Guillemot, J.W.T. Hessels, A. Jessner, C. Jordan, R. Karuppusamy, M. Kramer, A. Lassus, K. Lazaridis, K.J. Lee, K. Liu, A.G. Lyne, J. McKee, C.M.F. Mingarelli, D. Perrodin, A. Petiteau, A. Possenti, M.B. Purver, P.A. Rosado, S. Sanidas, A. Sesana, G. Shaifullah, R. Smits, S.R. Taylor, G. Theureau, C. Tiburzi, R. van Haasteren, A. Vecchio, Mon. Not. R. Astron. Soc. **458**, 3341 (2016). https://doi.org/10.1093/mnras/stw483

G.S. Downs, Interplanetary Navigation Using Pulsating Radio Sources. NASA Tech. Rep. 74N34150 (JPL Tech. Rep. 32-1594), Jet Propulsion Laboratory, Pasadena, CA, USA (1974)

C.M. Espinoza, A.G. Lyne, B.W. Stappers, M. Kramer, Mon. Not. R. Astron. Soc. **414**, 1679 (2011)

R.S. Foster, D.C. Backer, Astrophys. J. **361**, 300 (1990). https://doi.org/10.1086/169195

P.C.C. Freire, C.G. Bassa, N. Wex, I.H. Stairs, D.J. Champion, S.M. Ransom, P. Lazarus, V.M. Kaspi, J.W.T. Hessels, M. Kramer, J.M. Cordes, J.P.W. Verbiest, P. Podsiadlowski, D.J. Nice, J.S. Deneva, D.R. Lorimer, B.W. Stappers, M.A. McLaughlin, F. Camilo, Mon. Not. R. Astron. Soc. **412**, 2763 (2011)

P.C.C. Freire, N. Wex, G. Esposito-Farèse, J.P.W. Verbiest, M. Bailes, B.A. Jacoby, M. Kramer, I.H. Stairs, J. Antoniadis, G.H. Janssen, Mon. Not. R. Astron. Soc. **423**, 3328 (2012)

R.W. Hellings, G.S. Downs, Astrophys. J. **265**, L39 (1983)

G. Hobbs, W. Coles, R.N. Manchester, M.J. Keith, R.M. Shannon, D. Chen, M. Bailes, N.D.R. Bhat, S. Burke-Spolaor, D. Champion, A. Chaudhary, A. Hotan, J. Khoo, J. Kocz, Y. Levin, S. Oslowski, B. Preisig, V. Ravi, J.E. Reynolds, J. Sarkissian, W. van Straten, J.P.W. Verbiest, D. Yardley, X.P. You, Mon. Not. R. Astron. Soc. **427**, 2780 (2012)

A.W. Hotan, M. Bailes, S.M. Ord, Mon. Not. R. Astron. Soc. **369**, 1502 (2006)

V.M. Kaspi, M. Kramer, in *Astrophysics and Cosmology—Proceedings of the 26th Solvay Conference on Physics*, ed. by R. Blandford et al. (2016)

J. Kormendy, L.C. Ho, Annu. Rev. Astron. Astrophys. **51**, 511 (2013)

M. Kramer, Int. J. Mod. Phys. D **25**, 1630029-61 (2016). https://doi.org/10.1142/S0218271816300299

P.D. Lasky, C.M.F. Mingarelli, T.L. Smith, J.T. Giblin, E. Thrane, D.J. Reardon, R. Caldwell, M. Bailes, N.D.R. Bhat, S. Burke-Spolaor, S. Dai, J. Dempsey, G. Hobbs, M. Kerr, Y. Levin, R.N. Manchester, S.

Osłowski, V. Ravi, P.A. Rosado, R.M. Shannon, R. Spiewak, W. van Straten, L. Toomey, J. Wang, L. Wen, X. You, X. Zhu, Phys. Rev. X **6**(1), 011035 (2016). https://doi.org/10.1103/PhysRevX.6.011035

L. Lentati, S.R. Taylor, C.M.F. Mingarelli, A. Sesana, S.A. Sanidas, A. Vecchio, R.N. Caballero, K.J. Lee, R. van Haasteren, S. Babak, C.G. Bassa, P. Brem, M. Burgay, D.J. Champion, I. Cognard, G. Desvignes, J.R. Gair, L. Guillemot, J.W.T. Hessels, G.H. Janssen, R. Karuppusamy, M. Kramer, A. Lassus, P. Lazarus, K. Liu, S. Osłowski, D. Perrodin, A. Petiteau, A. Possenti, M.B. Purver, P.A. Rosado, R. Smits, B. Stappers, G. Theureau, C. Tiburzi, J.P.W. Verbiest, Mon. Not. R. Astron. Soc. **453**, 2576 (2015). https://doi.org/10.1093/mnras/stv1538

D.N. Matsakis, J.H. Taylor, T.M. Eubanks, Astron. Astrophys. **326**, 924 (1997)

M.A. McLaughlin, I.H. Stairs, V.M. Kaspi, D.R. Lorimer, M. Kramer, A.G. Lyne, R.N. Manchester, F. Camilo, G. Hobbs, A. Possenti, N. D'Amico, A.J. Faulkner, Astrophys. J. **591**, L135 (2003)

J.R. Oppenheimer, G. Volkoff, Phys. Rev. **55**, 374 (1939)

F. Özel, P. Freire, Annu. Rev. Astron. Astrophys. **54**, 401 (2016)

P.C. Peters, Phys. Rev. **136**, 1224 (1964)

D.J. Reardon, G. Hobbs, W. Coles, Y. Levin, M.J. Keith, M. Bailes, N.D.R. Bhat, S. Burke-Spolaor, S. Dai, M. Kerr, P.D. Lasky, R.N. Manchester, S. Osłowski, V. Ravi, R.M. Shannon, W. van Straten, L. Toomey, J. Wang, L. Wen, X.P. You, X.J. Zhu, Mon. Not. R. Astron. Soc. **455**, 1751 (2016). https://doi.org/10.1093/mnras/stv2395

J.E. Riedel, S. Bhaskaran, S. Desai, D. Han, B. Kennedy, G.W. Null, S.P. Synnott, T.C. Wang, R.A. Werner, E.B. Zamani, Autonomous Optical Navigation (AutoNav) DS1 Technology Validation Report. Deep Space 1 Technology Validation Reports (Rep. A01-26126 06-12), Jet Propulsion Laboratory, Pasadena, CA, USA (2000)

H.P. Robertson, Ann. Math. **38**, 101 (1938)

A. Sesana, A. Vecchio, Class. Quantum Gravity **27**(8), 084016 (2010)

A. Sesana, A. Vecchio, C.N. Colacino, Mon. Not. R. Astron. Soc. **390**, 192 (2008)

A. Sesana, F. Shankar, M. Bernardi, R.K. Sheth, Mon. Not. R. Astron. Soc. **463**, L6 (2016). https://doi.org/10.1093/mnrasl/slw139

F. Shankar, M. Bernardi, R.K. Sheth, L. Ferrarese, A.W. Graham, G. Savorgnan, V. Allevato, A. Marconi, R. Läsker, A. Lapi, Mon. Not. R. Astron. Soc. **460**, 3119 (2016)

R.M. Shannon, V. Ravi, L.T. Lentati, P.D. Lasky, G. Hobbs, M. Kerr, R.N. Manchester, W.A. Coles, Y. Levin, M. Bailes, N.D.R. Bhat, S. Burke-Spolaor, S. Dai, M.J. Keith, S. Osłowski, D.J. Reardon, W. van Straten, L. Toomey, J.B. Wang, L. Wen, J.S.B. Wyithe, X.J. Zhu, Science **349**, 1522 (2015). https://doi.org/10.1126/science.aab1910

I.I. Shapiro, Phys. Rev. Lett. **13**, 789 (1964)

R. Smits, M. Kramer, B. Stappers, D.R. Lorimer, J. Cordes, A. Faulkner, Astron. Astrophys. **493**, 1161 (2009)

T.M. Tauris, N. Langer, P. Podsiadlowski, Mon. Not. R. Astron. Soc. **451**, 2123 (2015)

J.H. Taylor Jr., Proc. IEEE **79**, 1054 (1991)

J.H. Taylor, Philos. Trans. R. Soc. Lond. A **341**, 117 (1992)

J.H. Taylor, in *Les Prix Nobel* (Norstedts Tryckeri, Stockholm, 1994), pp. 80–101

R. van Haasteren, Y. Levin, Mon. Not. R. Astron. Soc. **428**, 1147 (2013). https://doi.org/10.1093/mnras/sts097

J.P.W. Verbiest, M. Bailes, W. van Straten, G.B. Hobbs, R.T. Edwards, R.N. Manchester, N.D.R. Bhat, J.M. Sarkissian, B.A. Jacoby, S.R. Kulkarni, Astrophys. J. **679**, 675 (2008)

J.P.W. Verbiest, L. Lentati, G. Hobbs, R. van Haasteren, P.B. Demorest, G.H. Janssen, J.B. Wang, G. Desvignes, R.N. Caballero, M.J. Keith, D.J. Champion, Z. Arzoumanian, S. Babak, C.G. Bassa, N.D.R. Bhat, A. Brazier, P. Brem, M. Burgay, S. Burke-Spolaor, S.J. Chamberlin, S. Chatterjee, B. Christy, I. Cognard, J.M. Cordes, S. Dai, T. Dolch, J.A. Ellis, R.D. Ferdman, E. Fonseca, J.R. Gair, N.E. Garver-Daniels, P. Gentile, M.E. Gonzalez, E. Graikou, L. Guillemot, J.W.T. Hessels, G. Jones, R. Karuppusamy, M. Kerr, M. Kramer, M.T. Lam, P.D. Lasky, A. Lassus, P. Lazarus, T.J.W. Lazio, K.J. Lee, L. Levin, K. Liu, R.S. Lynch, A.G. Lyne, J. Mckee, M.A. McLaughlin, S.T. McWilliams, D.R. Madison, R.N. Manchester, C.M.F. Mingarelli, D.J. Nice, S. Osłowski, N.T. Palliyaguru, T.T. Pennucci, B.B.P. Perera, D. Perrodin, A. Possenti, A. Petiteau, S.M. Ransom, D. Reardon, P.A. Rosado, S.A. Sanidas, A. Sesana, G. Shaifullah, R.M. Shannon, X. Siemens, J. Simon, R. Smits, R. Spiewak, I.H. Stairs, B.W. Stappers, D.R. Stinebring, K. Stovall, J.K. Swiggum, S.R. Taylor, G. Theureau, C. Tiburzi, L. Toomey, M. Vallisneri, W. van Straten, A. Vecchio, Y. Wang, L. Wen, X.P. You, W.W. Zhu, X.J. Zhu, Mon. Not. R. Astron. Soc. **458**, 1267 (2016). https://doi.org/10.1093/mnras/stw347

J.M. Weisberg, D.J. Nice, J.H. Taylor, Astrophys. J. **722**, 1030 (2010)

N. Wex, arXiv:1402.5594 (2014)

C.M. Will, Living Rev. Relativ. **17**, 4 (2014)

K. Yagi, D. Blas, E. Barausse, N. Yunes, Phys. Rev. D **89**(8), 084067 (2014)

M.D. Young, R.N. Manchester, S. Johnston, Nature **400**, 848 (1999)
M. Yu, R.N. Manchester, G. Hobbs, S. Johnston, V.M. Kaspi, M. Keith, A.G. Lyne, G.J. Qiao, V. Ravi, J.M. Sarkissian, R. Shannon, R.X. Xu, Mon. Not. R. Astron. Soc. **429**, 688 (2013)
X.J. Zhu, G. Hobbs, L. Wen, W.A. Coles, J.B. Wang, R.M. Shannon, R.N. Manchester, M. Bailes, N.D.R. Bhat, S. Burke-Spolaor, S. Dai, M.J. Keith, M. Kerr, Y. Levin, D.R. Madison, S. Osłowski, V. Ravi, L. Toomey, W. van Straten, Mon. Not. R. Astron. Soc. **444**, 3709 (2014). https://doi.org/10.1093/mnras/stu1717

Space Sci Rev (2018) 214:66
https://doi.org/10.1007/s11214-018-0498-1

Very Long Baseline Interferometry: Dependencies on Frequency Stability

Axel Nothnagel[1] · Tobias Nilsson[2] · Harald Schuh[2]

Received: 18 November 2016 / Accepted: 28 March 2018 / Published online: 5 April 2018

Abstract Very Long Baseline Interferometry (VLBI) is a differential technique observing radiation of compact extra-galactic radio sources with pairs of radio telescopes. For these observations, the frequency standards at the telescopes need to have very high stability. In this article we discuss why this is, and we investigate exactly how precise the frequency standards need to be. Four areas where good clock performance is needed are considered: coherence, geodetic parameter estimation, correlator synchronization, and UT1 determination. We show that in order to ensure the highest accuracy of VLBI, stability similar to that of a hydrogen maser is needed for time-scales up to a few hours. In the article, we are considering both traditional VLBI where extra-galactic radio sources are observed, as well as observation of man-made artificial radio sources emitted by satellites or spacecrafts.

Keywords VLBI · Technical realization · Station clock requirements · Artificial signal generation

1 Introduction

High-precision geodetic and astrometric very long baseline interferometry (VLBI) has been used since the end of the 1970s. Pairs of radio telescopes as far apart as an Earth diame-

High Performance Clocks with Special Emphasis on Geodesy and Geophysics and Applications to Other Bodies of the Solar System
Edited by Rafael Rodrigo, Véronique Dehant, Leonid Gurvits, Michael Kramer, Ryan Park, Peter Wolf and John Zarnecki

✉ A. Nothnagel
nothnagel@uni-bonn.de

T. Nilsson
nilsson@gfz-potsdam.de

H. Schuh
schuh@gfz-potsdam.de

1 Institute of Geodesy and Geoinformation, University of Bonn, Nussallee 17, 53115 Bonn, Germany

2 GFZ German Research Centre for Geosciences, Telegrafenberg A17, 14473 Potsdam, Germany

ter simultaneously receive the electromagnetic radiation of extra-galactic radio sources like quasars and other compact radio galaxies at quasi-infinite distances. By cross-correlating the signals received by two telescopes, the difference in arrival time (denoted as the delay) can be inferred. By analyzing these delays, many important geodetic and astrometric parameters can be estimated, such as the positions of the telescopes and the radio sources, as well as the Earth Orientation Parameters (EOP). VLBI thus makes important contributions to the International Terrestrial Reference Frame (ITRF), is the sole technique used for realizing the International Celestial Reference Frame (ICRF), and is the only space geodetic technique able to estimate all five EOP (unique for UT1-UTC and precession/nutation).

The VLBI technique has also been used for tracking interplanetary spacecraft and for determining their trajectories with other methods than Doppler ranging in several projects. Quite recently, observations of near-Earth objects, such as artificial satellites, spacecraft in orbit around the Moon or a lander on the Moon, have come into the focus of geodesists for linking specific reference frames through common targets in space. With these near-field targets, the concept of geodetic VLBI has to be refined taking care of peculiarities of man-made sources transmitting radio signals, which are not positioned at quasi-infinity and which are not of natural origin.

In order for the VLBI technique to work, the stations need to be equipped with very precise clocks. In this article we review the importance of clocks in VLBI and discuss what accuracy is needed in different parts of the VLBI correlation and analysis chain. We begin in Sect. 2 with an overview of the VLBI technique, as it is used for observations of extra-galactic radio sources. In Sect. 2.1 we discuss what areas time and frequency stability are important. To further investigate the precision needed in VLBI data analysis, we present the results of a few simulations in Sect. 2.2. The rest of the paper is devoted to the observation of artificial radio sources from near-field targets with VLBI, beginning with a general description in Sect. 3. The generation of the artificial signals is discussed in Sect. 3.1, and requirements of the clocks in Sect. 3.2.

2 General VLBI Concepts

Quasars and other compact extra-galactic radio sources are found at the frontiers of the universe at quasi-infinite distances. Ideal candidates of these sources are point-like and show no proper motions on decadal time scales. Due to their distances, the emission of these radio sources arrives on Earth as plain wavefronts and have different arrival times at the two radio telescopes forming a baseline (Fig. 1). The difference in arrival times, T_i, is the delay, τ_{obs}, which is the primary observable of geodetic and astrometric VLBI:

$$\tau_{obs} = T_2 - T_1. \tag{1}$$

In order to obtain the delay, several steps are needed. First, the telescopes (two or more) observe the radio source simultaneously. Since the telescopes are highly directional (beamwidth of a few hundred arcseconds), practically only one radio source can be observed at a time. The signal from the radio-source is normally observed at two frequency bands, S-band (2.2–2.4 GHz) and X-band (8.2–8.9 GHz); in the future this may be extended to four bands between 2 GHz and 14–18 GHz. Once the signal is received at a station, it is mixed with a reference signal generated at the station (see Fig. 1) in order to covert it down to a few hundred MHz. Then the signal is recorded, and given time-tags using the time from the station clock. The recorded data are then transferred to a correlator. Traditionally this has

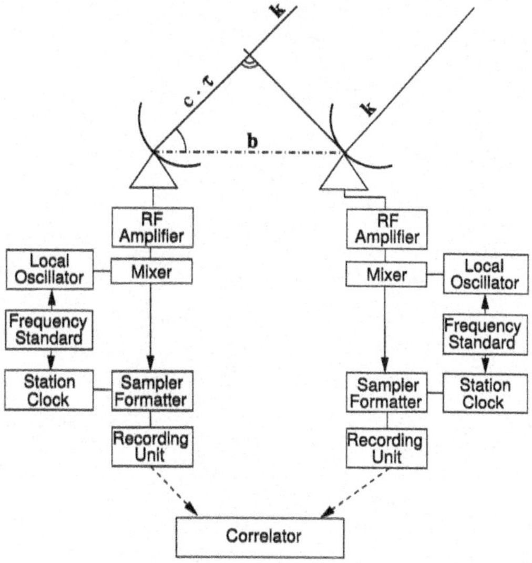

Fig. 1 VLBI constellation with electronics components

been done by shipping the recording media, however in recent years electronic transfer has been becoming more and more common. At the correlator, the recorded signals from the different telescopes are cross-correlated. By analyzing the output from this cross-correlation, the VLBI observables are obtained (for details, see e.g. Thompson et al. 2007): the phase, the group delay (the derivative of the phase w.r.t. frequency) and the phase rate (the time derivative of the phase). Of these three observables, normally only the group delay is used in the further analysis. In principle, the phase delays (which can be obtained from the phases) would be more accurate, however, these contain ambiguities of unknown multiples of the signal wavelengths (3.6 cm at 8.4 GHz). With the current VLBI system, these ambiguities are very difficult to resolve, although there is great hope that this will be possible in the future VLBI system, VGOS (VLBI Geodetic Observing System).

The precision of the group delay is given by Whitney (1974):

$$\sigma_\tau = \frac{1}{2\pi SNR \nu_{rms}} \tag{2}$$

where SNR denotes the signal-to-noise-ratio, $\nu_{rms} = \sqrt{\langle\nu^2\rangle - \langle\nu\rangle^2}$ is the root-mean-square of the observed bandwidth, and ν denotes the frequency. From the formula it is obvious that in order to obtain high precision, a large bandwidth should be observed. Thus, to avoid having to record at a very high data rate, the frequency band is sampled in discrete frequency channels (typically 8 in X-band and 6 in S-band). This keeps the recorded bandwidth down, and at the same time allows for a large ν_{rms} and thus a high precision of the group delays. This procedure is called bandwidth synthesis (Rogers 1970).

Normally, calibrations need to be performed in order to correct the observed delays for the delay in the receiving system between the feed horn and the formatter (where the time-tagging and recording is performed), see Fig. 2. Two calibrations have to be distinguished here. The first one is a phase calibration (phase-cal) which is done to measure the delay between the feed-horn and the formatter in each channel. This is important to make since the delay in the receiving system can be dispersive, hence different channels will be affected differently and can thus cause a problem when estimating the group delay (i.e. the slope

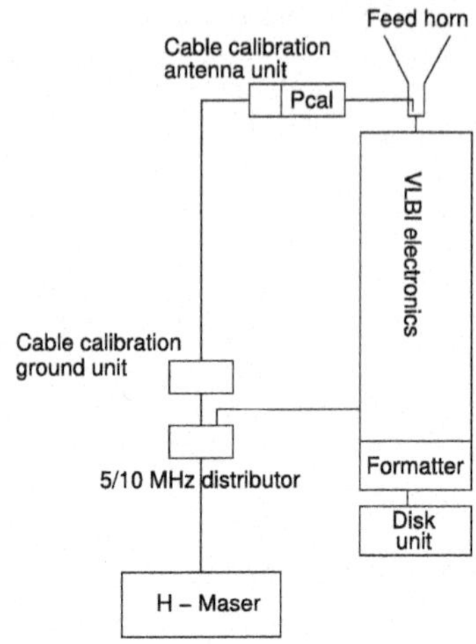

Fig. 2 VLBI delay and cable calibration system

of the phase of the different channels w.r.t. frequency) if not calibrated. This calibration is based on a series of monochromatic tones with 1 MHz separations generated in the Pcal unit in Fig. 2, using frequency information from the station clock. These tones are practically identical to a sequence of pulses separated by 1 μs. These are fed into the feed horn, where all tones are in phase, and run through the system together with the quasar signals. The phases of the phase-cal signals are then extracted at time of correlation and used to calibrate the delay observables.

Another calibration is the cable delay calibration (cable-cal) which in fact is the calibration of the trigger signal for the phase-cal tones. The generation of the phase-cal tones is driven by a 5 MHz signal from the station clock which is brought up from the control building to the location of the phase-cal system near the feed horn in the radio telescope (Fig. 2). The electrical length of the cable used for this can vary due to cable stretching and thermal effects, thus it needs to be calibrated. This calibration is done by a two-way phase measurement where a 5 MHz signal is sent up from the control room to the front end. This is then modulated by a 5 kHz signal to produce a 5.005 MHz signal, which is returned to the control room. The latter is mixed with the original 5 MHz signal, and then low-pass filtered to produce a 5 kHz signal. This is then compared to a locally generated 5 kHz signal, which gives a measure of the travel time between the control room and the front-end.

The observed group delay will depend on a number of different terms, as given by the following equation:

$$\tau_{obs} = \tau_{geom} + \tau_{iono} + \tau_{atm} + \tau_{clk} + \tau_{instr} + \tau_{def} + \epsilon. \tag{3}$$

The geometrical delay, τ_{geom}, is normally the most interesting part for geodesists since it is dependent on the relative position of the telescopes, the orientation of the Earth, and the positions of the radio sources. The ionosphere delay, τ_{iono}, and the neutral atmosphere delay, τ_{atm}, are caused by the fact that the propagation speed of radio waves in these media is different from that of vacuum, as well as the increased propagation distance due to bending. The ionospheric delay is frequency dependent ($\tau_{iono} \propto \nu^{-2}$), thus by making a linear combination of the delays observed at two (or more) frequencies, this effect can be removed (this is the reason for observing two frequency bands). The delay in the neutral atmosphere, however, is frequency independent, thus this part needs to be modeled in the data analysis. The contribution to the delay caused by the difference in rate of the independent reference oscillators (clocks) is contained in the term, τ_{clk}, while the (uncalibrated) delays in the receiving systems are contained in the term τ_{instr}. If phase-cal and cable-cal are both properly applied, τ_{instr} should be small, although some uncalibrated effects still remain (Fig. 2). In any case, the behavior of τ_{instr} is in principle indistinguishable from variations in the station clocks (τ_{clk}), thus these two terms are practically treated together and it can be said that τ_{instr} contribute to the total clock error budget. The term τ_{def} contain the effects of deformations of the VLBI antennas, e.g., thermal and gravitational deformations. Finally, ϵ denotes the observation noise.

Once the delays τ_{obs} have been determined, these are analyzed to determined the parameters of interest, such as the station and radio source coordinates, as well as the EOP. Furthermore, since many of the effects contributing to the delay (Eq. (3))—such as the atmosphere and the clock errors—are not known precisely a priori, these also need to be parametrized and estimated in the analysis. For the parameter estimation, several different methods can be used, like Kalman Filter (KF) (Nilsson et al. 2015) and Least-Squares Collocation (LSC) (Titov 2000). However, we here limit the discussion to the classical least squares method (LSQ), which is the most commonly used method. Furthermore, since this article is about clocks, we only discuss the modeling of the station clocks. It should be noted here that since the VLBI telescopes can only observe one radio source at a time, the temporal variations of the station clocks must be modeled. This is different from, e.g., GNSS (Global Navigation Satellite Systems) analysis, where observations to many satellites are typically available at each epoch, thus allowing the clock errors to be estimated epoch-wise (or differenced away by forming double differences). As will be discussed, this also means that the stability of the clocks used in VLBI need to be much better than those commonly used for GNSS.

For the VLBI station clocks, the following parameterization is commonly used:

$$\begin{aligned}\tau_{clk}(t) &= \left(\mathcal{T}_0^B + \mathcal{T}_1^B \cdot (t - t_0) + \mathcal{T}_2^B \cdot (t - t_0)^2\right) \\ &\quad - \left(\mathcal{T}_0^A + \mathcal{T}_1^A \cdot (t - t_0) + \mathcal{T}_2^A \cdot (t - t_0)^2\right) \\ &\quad + t^B(t_i)\frac{t - t_{i-1}}{t_i - t_{i-1}} + T^B(t_{i-1})\frac{t_i - t}{t_i - t_{i-1}} \\ &\quad + T^A(t_i)\frac{t - t_{i-1}}{t_i - t_{i-1}} + T^A(t_{i-1})\frac{t_i - t}{t_i - t_{i-1}} \end{aligned} \tag{4}$$

where t denotes the time and $t_{i-1} < t < t_i$, and $\mathcal{T}_i^j$ and T_i^j are unknown coefficients to be estimated in the least squares inversion. A and B refer to the first and second station, respectively, of the baseline observing the delay. The first two lines of Eq. (4) describe the average variations of the clock over one day, which can be approximated by a second degree polynomial. The other part of the equation describes the short term variations of the clocks, which are normally modeled as piece-wise linear functions. The typical resolution of these piece-wise functions (i.e. $t_i - t_{i-1}$) are normally of the order 20 min to 1 h.

More background information and detailed explanations on standard VLBI applications can be found in several text books or review articles such as Sovers et al. (1998) and the many references therein. The future of geodetic VLBI will be the VLBI Global Observing System (VGOS) which will employ smaller, more agile telescopes and a much larger recorded bandwidth for precision enhancement. For details see Petrachenko et al. (2009) and Petrachenko et al. (2012).

2.1 Clock and Frequency Standard Stability

As seen above, there are several places in the VLBI observation/analysis chain where precise time and/or frequency information is needed. In this sub-section, we discuss these in more detail, and assess what clock precision is needed. We here consider four different areas:

- The first item refers to coherence within the time period of each individual scan (tracking of a radio source with two or more radio telescopes), which, in most cases, can last between 10 s and 300 s. For the correlation process to work well, the signals recorded by the two telescopes need to be coherent, otherwise the SNR will go down leading to increased noise in the delay (see Eq. (2)) or, in the worst case, that the correlation process is unsuccessful and no delay can be found. However, the two signals will not be perfectly coherent because of random phase fluctuations in the recorded signals caused by, among other effects, instabilities in the station clocks (since these are used for generating the reference frequency for the mixer as well as controlling the sampler). This sets a limit on the required precision of the station clocks. One simple equation for calculating the needed precision is (Thompson et al. 2007):

  $$2\pi \nu_0 T \sigma_c(T) \leq 1. \tag{5}$$

  This equation gives the limit where the RMS phase error is less than 1 radian. The frequency is denoted by ν_0, the integration time T, and $\sigma_c(T)$ is the Allan standard deviation (ASD) of the clock over the time period T. Assuming an integration time of 300 s, which is about the period of the longest scans for the current geodetic VLBI system, and the highest observing frequency ν_0 of 8.4 GHz, gives a limit on the clock ASD of $\sigma_c < 6.3 \cdot 10^{-14}$ @ 300 s. For the future VGOS system, where the scans will be shorter (10 s) but the maximum frequency higher (maybe going up to 18 GHz), this limit change to $\sigma_c < 8.8 \cdot 10^{-13}$ @ 10 s.

  More precisely, the loss in SNR due to incoherence is given by the following equation (Thompson et al. 2007):

  $$SNR = \left| \frac{1}{T} \int_0^T e^{i\phi(t)} dt \right| SNR_0 = C(T)\, SNR_0 \tag{6}$$

  where SNR_0 is the SNR when the signals are perfectly coherent, and ϕ denote the phase fluctuations. If we assume that the clock and frequency noise can be considered white, it can be shown that (see Thompson et al. 2007 and Rogers and Moran 1981 for details):

  $$\sqrt{\langle C(T)^2 \rangle} = e^{-\frac{2\pi^2 \nu_0^2 \sigma_c^2 T^2}{3}} \tag{7}$$

  If we require that $\sqrt{\langle C(T)^2 \rangle} < 0.8$ (SNR loss less than 1 dB), this gives a requirement for the clock ASD of $\sigma_c < 7.3 \cdot 10^{-14}$ @ 300 s for the current VLBI system and $\sigma_c < 1.0 \cdot 10^{-12}$ @ 10 s for the future VGOS system. If the limit is set to $\sqrt{\langle C(T)^2 \rangle} < 0.5$ (SNR loss less than 3 dB), we get $\sigma_c < 1.3 \cdot 10^{-13}$ @ 300 s for the current VLBI system and $\sigma_c < 1.8 \cdot 10^{-12}$ @ 10 s for VGOS.

  The limits derived above put strong limits on what types of clocks that can be used for VLBI. A hydrogen maser (see Fig. 3) easily fulfills the limits, both for the current system and for VGOS. Thus, coherence is not a big problem since today all VLBI stations are using hydrogen masers. In principle, clocks with worse short-time stability, such as very good rubidium and cesium clocks, might work, however they are on the limit of fulfilling

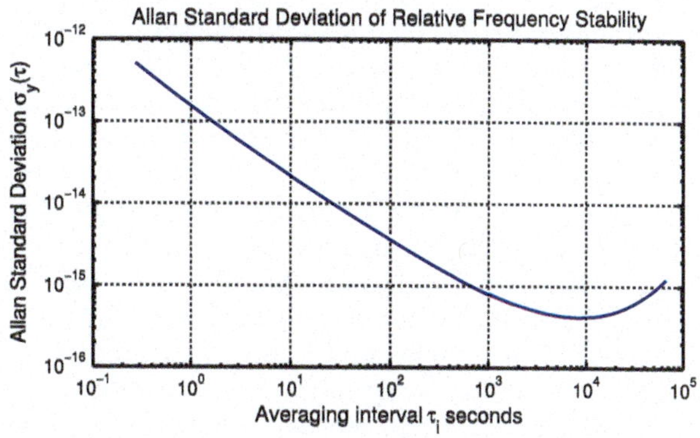

Fig. 3 Generic Allan standard deviation of hydrogen maser frequency standard

the requirements. Hence, it is not recommended to use such clocks for geodetic VLBI, neither for the current system nor for VGOS, however, such clocks might work as backup in cases where there are problems with the hydrogen maser.

- The second issue is the stability needed to link consecutive scans of a geodetic VLBI observing session. Because, as stated before, a VLBI telescope can only observe in one direction at a time, the station clock errors need to be parameterized (cf. Eq. (4)) and the clock parameters solved in the least squares adjustment. The fitted clock function needs to be able to approximate the actual behavior of the clock well, otherwise it is possible that errors are introduced in the other estimated parameters, e.g. the station coordinates. This can of course be achieved by selecting the estimation interval of the piece-wise linear functions short enough. On the other hand, this interval should not be too short. The estimation interval needs to be long enough so that the observations within each interval provide a good sky coverage, thus allowing for a good de-correlation of the estimated parameters, e.g. the atmospheric parameters, the clock and the coordinates. Currently, 5 to 15 scans per hour are scheduled by each of the telescopes in a network. When VGOS will be operational with its fast slewing telescopes this number will increase to 60–120 scans per hour. The estimation interval of the piece-wise linear clock functions are typically of the order 20 min to 1 h. As are found by the simulations in Sect. 2.2, this requires a clock precision similar to that of a hydrogen maser.
- For the correlation process itself, time keeping or synchronization of the two clocks of an interferometer driven by hydrogen masers is needed to the level of about 1 μs. This helps to limit the correlation process to time lags of a few microseconds before and after the expected delay. For this purpose, each VLBI station maintains GPS time transfer equipment which is used to regularly determine the deviations from UTC(GPS) to a few hundred nanoseconds. This appears to be very rough. However, there is no need to be more accurate because these numbers are only used to search for initial correlation maxima of a few selected scans within an observing session for determining a relative offset and a drift in the station clocks. The precise relative clock behavior is then determined a posteriori in the analysis of all observed group delays in a least-squares adjustment down to the picosecond level.

In cases where the a priori clock information is unknown or inaccurate, the correlation maximum will be outside of the limited trial lags. In such a case, the initial correlation will only produce noise and no significant peak in the delay resolution function. A repetition of the correlation is then necessary with a much wider search window. Again this will have to be done for a few observations within a 24 h period to determine the relative clock parameters up to a second order polynomial.

- The time tags of the recorded VLBI signals of a reference telescope normally determine the epochs of the observed delays $\tau(t)$. Later on, i.e., in the subsequent analysis these epochs are relevant for the time reference of the target parameters $UT1\text{-}UTC$. As UTC is actually taken from the observation epochs, any error in the clock of the reference telescope enters the $UT1\text{-}UTC$ parameters directly. For this reason, care is taken that biases in the station clocks are taken into account as much as possible. These are applied already in the setup of the correlation process where the time tags of each telescope are corrected for an empirical bias. For this reason, the $UT1\text{-}UTC$ time series of the IVS is self consistent. However, these biases are only relative to the radio telescope of Kokee Park on Hawaii run by the US Naval Observatory which by itself is one of the major laboratories for producing Temps Atomique International (TAI). Although any remaining bias with respect to UTC proper is believed to be not larger than 1 μs, this uncertainty in the timing domain still exists (Himwich et al. 2017).

2.2 Simulations

One way to assess the impact of the station clock errors on the estimated parameters, e.g., the station positions, is to perform simulations. Here we present results of such simulations for two different VLBI networks. One is the network of the CONT14 campaign, which was a 15 day long continuous campaign performed in May 2014 and can be considered as the state-of-the-art of the current VLBI system. For the simulations, we assumed the same observing schedules as were used for the actual observations. The other network was a potential future 30-station VGOS network. Here we generated an observing schedule using the VIE_SCHED software (Sun et al. 2014), assuming that all stations were of VGOS standard and equipped with broadband receivers. Due to rapid source switching, which is the goal of VGOS, a total of 181817 delay observations in 24 h have been used in the simulations.

We simulated VLBI group delays using the procedure presented in Pany et al. (2011). This procedure generates delays affected by the three most important random error sources in geodetic VLBI: tropospheric delays, clock errors, and observation noise. The tropospheric delays were simulated using the algorithms presented in Nilsson and Haas (2010), the observation noise was simulated as white noise (30 ps standard deviation for the CONT14 simulations and 10 ps for the VGOS simulation), and each clock error as the sum of an integrated random walk processes and a random walk process. Simulations were done assuming different Allan standard deviations (ASD) for the clocks, between 10^{-16} and 10^{-13} @ 50 min. The simulated delays were then used as input to a standard VLBI analysis. The simulated data from each day were analyzed individually, i.e., we obtained one set of station positions per day.

Figure 4 shows the median station position repeatabilities obtained from the CONT14 simulations, as function of the assumed ASD of the clocks. In Fig. 5 the corresponding repeatabilities from the simulations with the VGOS network are demonstrated. As can be seen, the repeatabilities get better as more stable clocks are assumed (lower ASD). However, there is hardly any improvement using clocks with a stability better than 10^{-14} @ 50 min. The reason is that other error sources, most importantly the troposphere, get more significant than the clock errors at this point. Hence we can conclude that, for VLBI data analysis, the clocks with a stability of 10^{-14} @ 50 min are sufficient, what is effectively the performance of the clocks currently operated at the VLBI stations (see, e.g., Rieck et al. 2012). Similar results were found by e.g., Petrachenko et al. (2009) and Pany et al. (2011).

To further test the limit the clocks put on the precision of the estimated parameters, we also made simulations with the VGOS network where the atmospheric fluctuations were

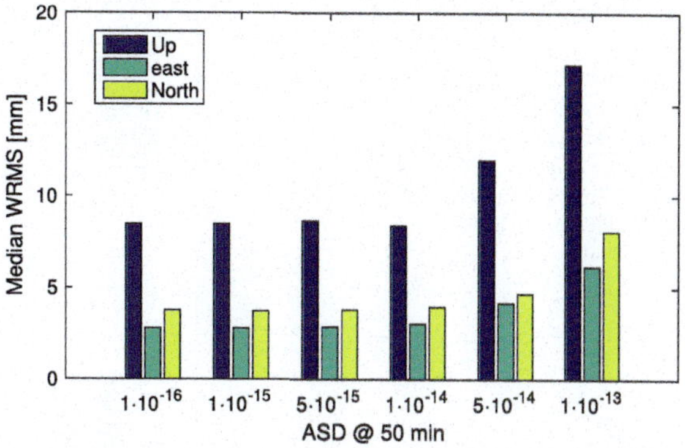

Fig. 4 Median station position repeatability for the CONT14 simulations

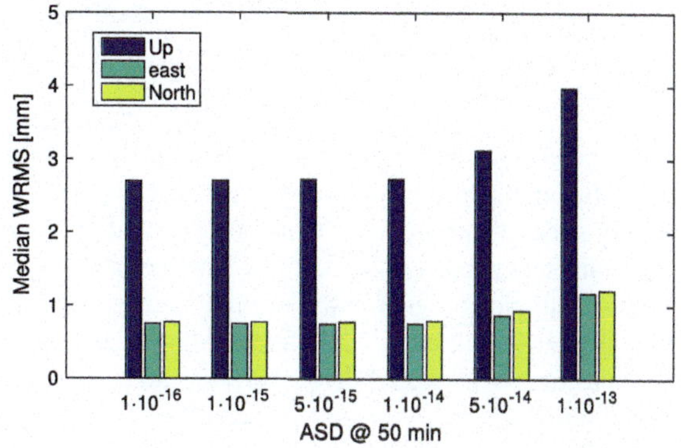

Fig. 5 Median station position repeatability for the simulations using the 30 station VGOS network and 181817 delay observations

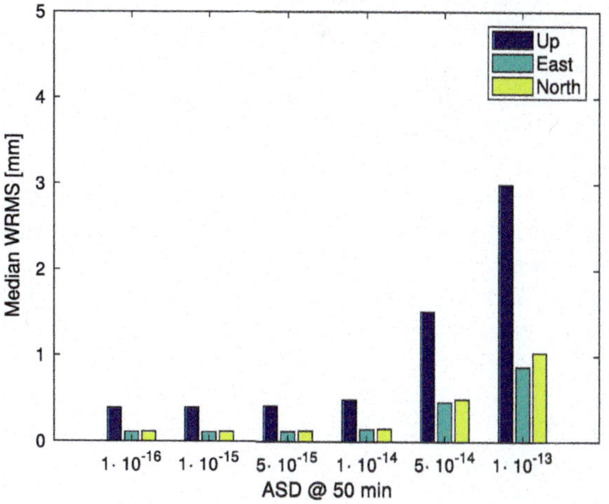

Fig. 6 Median station position repeatability for the simulations using the 30 station VGOS network and 181817 delay observations. The fluctuations in the atmosphere were reduced by a factor 10, compared to Fig. 5

reduced by a factor of 10. The results for the station position repeatabilities can be seen in Fig. 6. By comparing with Fig. 5 we can clearly see how much the atmosphere limits the precision. Furthermore, we can also note that if the atmosphere would not be such a problem, then there would be an improvement when using more accurate clocks, like clocks with ASD of 10^{-15} @ 50 min. Thus, for VLBI telescopes located in places with a very stable atmosphere, e.g. at 4–5 km altitude, using better clocks would be beneficial.

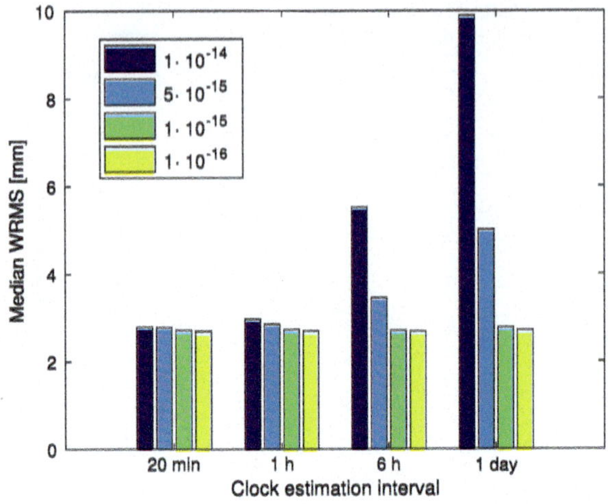

Fig. 7 Median station position repeatability for the simulations using the 30 station VGOS network and 181817 delay observations. Shown are the results when using different estimation intervals as well as for different clock ASDs

Another issue needed to be considered is the estimation interval of the piece-wise linear clock functions. In the solutions presented in Fig. 5 an interval of 20 min was used. However, if very good clocks are used, this interval could be increased, which will reduce the number of parameters needed to be estimated and could thus increase the precision. To test this, we made simulations (again with "normal" atmosphere) where the estimation interval for the clocks were increased to 1 h, 6 h, or 24 h. The results can be seen in Fig. 7. We can see that when using clocks with ASD of 10^{-15} @ 50 min the estimation interval could be increased to 6 h or even 1 day without significant loss in precision. This, however, does not give any noticeable improvement in the station position repeatabilities. The likely reason is that the coordinates are estimated as constant over the whole session, i.e. 1 day, thus most important is the mean offset of the clock over the day, which is estimated in all cases. It is possible that improvements could be seen if the station coordinates were estimated with a higher resolution, or if the clocks were estimated in even longer intervals (using data from several sessions, but keeping the daily constitution of the station coordinates). This would, however, require more investigations, which is beyond the scope of this review article.

It should be mentioned here that of course these simulations only consider the pure hydrogen maser stability. Any error in the instrumental delay as elaborated on in Sect. 2.1 have not been considered here. Likewise, the assumptions made here are based on the current status of atmosphere modeling and on telescope locations with medium range atmospheric water vapor and beyond. However, in Pany et al. (2011) it was shown that in simulations the difference between telescopes caused by the refractive index structure constant C_n can be more than 50% larger than that of the hydrogen maser clock with an ASD of $\leq 1 \times 10^{-14}$ at 50 min. From this we can conclude that the current clock stability will be sufficient for some time to come.

3 VLBI of Nearfield Targets

With VLBI it is also possible to observe artificial signal sources transmitted by satellites or spacecrafts. For example, VLBI has long been used for tracking of interplanetary spacecrafts (Border 2009; Duev et al. 2016). In recent years, there has also been an increased interest of using VLBI to observe satellites orbiting the Earth (Tornatore et al. 2014; Plank et al.

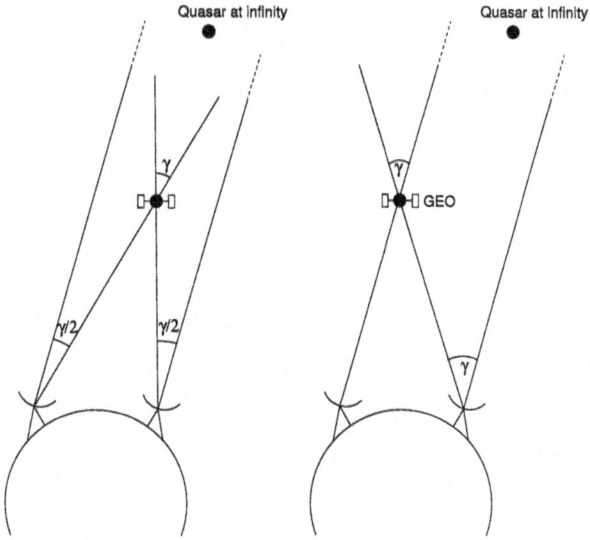

Fig. 8 Pointing directions of radio telescopes depending on satellite altitude

2014; Anderson et al. 2018) as well as orbiters around and landers on the moon (Tang et al. 2014). One important reason for observing satellites with VLBI is the possibility to have co-location of several space geodetic techniques in space. If one satellite can be observed by several space geodetic techniques, and the relative positions of the different techniques on the satellite are accurately known, such observation can be used as a tie between the reference frames realized between the individual techniques. It would thus be an alternative or a compliment to having local ties measured on the ground, and would help to improve multi-technique reference frames such as the ITRF (International Terrestrial Reference Frame, Altamimi et al. 2016). Thus a lot of effort has been applied to the observing GNSS satellites with VLBI (Tornatore et al. 2014; Plank et al. 2015), and there has been proposals for satellite missions having all the four main space geodetic techniques co-located, e.g. GRASP (Bar-Sever et al. 2009) and E-GRASP (Biancale et al. 2017). It should be noted that VLBI of near-field targets, especially for observation of satellites orbiting the Earth, is still a developing technique. Thus, there are still possibilities for technological improvements, and optimal observing and analysis strategies need to be developed. Nevertheless, we here give a basic outline of observing near-field targets and discuss the requirements of the clocks used at the stations and on-board the satellites.

There are several important differences between observing quasars and observing artificial signals which we need to consider. First of all, there are technical differences which may cause limitations. For example, in particular when observing a signal not designed for VLBI, e.g., when observing a GNSS satellite, the signal strength will be many magnitudes stronger than that of a quasar, the frequency will be different from the normal VLBI frequencies, and the bandwidth of the signal will be limited. Furthermore, the satellite can move relatively fast, making it more difficult to track. This is particularly an issue for Low Earth Orbit Satellites (LEO). For LEO satellites it is also difficult to have a common visibility of the satellite from two stations, in particular for long baselines (see Fig. 8).

For observing artificial near-field targets, we can identify three different strategies:

- The first strategy is satellite VLBI mode. This means that the observations are done similar to normal quasar observations. This method is suitable if some observations of a satellite are to be included in a normal VLBI schedule observing quasars. As mentioned

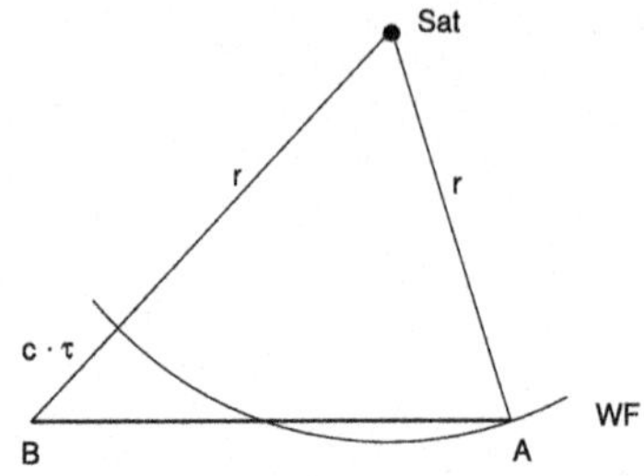

Fig. 9 Near-field geometry with wavefront (WF) arriving at telescope A and producing a delay τ before it arrives at telescope B

above, it is important to have short baselines in order to have common visibility of the satellite from two (or more) stations.

- The second strategy is differential VLBI (or phase referencing). This means alternating between observing a spacecraft and quasars located in approximately the same directions. Since many of the error sources will be similar for the spacecraft and the quasars, these can approximately be removed by taking the difference. This technique is suitable for determining the orbit or position of the spacecraft in the celestial reference frame, i.e. it is commonly used when tracking interplanetary spacecraft. However, since it requires that at least two stations can see the satellite, and that there are also a quasar nearby as seen from both stations, it is not very useful for observations of LEO satellites.
- The third strategy is Time of Flight (TOF) measurements. Here the artificial signals have a known modulation, allowing for range measurements to be made to the satellite using just one VLBI telescope (e.g., similar to how GNSS work). Since with this strategy we only need to observe the satellite with one VLBI telescope, this is advantageous especially for LEO orbits. However, the satellite needs to be equipped with an atomic frequency standard for the signal generation. It should be noted that, strictly speaking, this mode is not VLBI since nothing is interfered. However, since it uses VLBI telescopes it is worth discussing it in the context of VLBI satellite observations.

The big difference between the observed delay from a near-field target and from a quasar at quasi-infinite distance is that for near-field targets the wave-front is curved, see Fig. 9. See, e.g. Duev et al. (2012) for explicit expression of the geometrical delay. Otherwise, the same terms as given in Eq. (3) need to be taken into account also here. In addition, if TOF measurements are used, we also need to consider the error in the satellite clock. Furthermore, in some cases the observation of the satellite is done only in one frequency band, thus the ionosphere needs to be calibrated in another way. For example, some of the test observations of GNSS satellites, that have been done, were only able to observe the higher of the two GNSS frequencies. Normally, in these cases the ionospheric calibration is done by using global total electron content maps.

If satellite VLBI mode or TOF mode is used to track the satellite, the parameter estimation is similar to the case of quasar observations. The difference is that the satellite positions—and for TOF also the satellite clocks—need to be parameterized and estimated. Typically, the position is parameterized as offsets in radial, along-track, cross-track, or as orbit parameters. In general the highest sensitivity of traditional VLBI is to the along- and cross-track components of the position, while the radial is not as well determined. This is particularly true for higher orbits, lunar orbiters, and interplanetary spacecrafts. If TOF measurements are used, the radial component could in principle be better constrained, although there will be a high correlation with the satellite clock offset. For the modeling of the satellite clock offset, the same parameterization as for the station clocks (Eq. (4)) can be used.

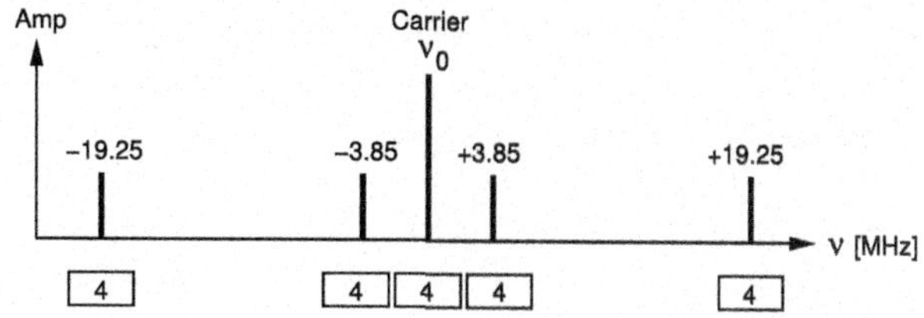

Fig. 10 Example of differential one-way ranging (DOR) tone allocation

3.1 Artificial Signal Sources

In traditional VLBI, the signals observed are the random emission from quasars. Thus, for typical IVS observations, the satellite signal generator does not need to be more complicated than a simple noise generator, as long as the emission is rather uniform over the frequency bands recorded by the VLBI telescopes. If normal quasar observations are to be done along with the observations of the artificial target, it is important that the satellite signal power is not too many orders of magnitude stronger than the power of the quasars. Otherwise, there is a risk that the VLBI receivers will saturate.

In many cases spacecraft transmit so-called DOR tones (Differential one-way ranging). They consist of a central carrier and mostly two tones below and above the carrier frequency (Fig. 10). Space agencies observe these tones mostly in narrow bands of a few hundred kHz and often with up to 8-bit sampling.

VLBI receiver chains are usually configured for observing bands of several MHz (2, 4, 8, 16 and even up to 32 MHz). This is to compensate for the extremely weak signal levels received from extra-galactic radio sources. Since the stations are reluctant to change these setups, observations of these DOR tones for ΔDOR applications can often only be carried out in the wide-band configuration. Unfortunately, however, bandpass effects are a critical issue which are investigated at present.

3.2 Requirements for Clock and Frequency Standard Stability

For the correlation of artificial signals, in principle the same requirements w.r.t coherence apply as for quasar signals (Sect. 2.1). For the data analysis, the requirements on the clock precision is dependent on the observation and analysis strategy applied. Since these are partly still being developed, we cannot at present give any definite statement on the clock precision, however, we can give a few rough estimates.

- For satellite VLBI mode, we would expect that there would be little or no difference to the case when observing only quasars, thus the limits obtained in Sect. 2.2 will be valid also here. For this strategy, the precision of the satellite clock is unimportant.
- In differential VLBI, the difference between the spacecraft observation and a reference quasar is used, thus approximately removing many error sources, including the station clocks. Plank (2013) investigated how well different error sources were really canceled. Assuming a switching interval of 60 s, the most important remaining error was found to be the atmospheric delay (RMS error of 25 ps). If we require the remaining clock error to be less than this, this would require a station clocks with ASD lower than $4 \cdot 10^{-13}$ @ 60 s. A hydrogen maser fulfills this requirement; however, rubidium or cesium clocks would not.
- The TOF strategy is the least developed of the three strategies, thus the requirements for this strategy is most uncertain. This is especially true if we consider multi-technique satellites such as GRASP or E-GRASP. For such satellites, observations from other space

geodetic techniques could also be available, which could help constraining the satellite position and the satellite clock. However, in any case we could expect that the requirements for the station clocks would not be much different form traditional VLBI. For TOF we also need to consider the satellite clocks. If the satellite clock are to be determined from only the VLBI observations, we would expect that the requirements for these clocks should be on the same order as the station clocks. However, to obtain a definite limit, further investigations are needed. If additional information about these clocks are available, e.g., determined from GNSS measurements, less precise satellite clocks could potentially also be used. However, in that case it should be pointed out that the VLBI results will no longer be independent of the GNSS results.

If a satellite is observed by multiple space geodetic techniques, and the orbit of the spacecraft is to be determined in a combined solution using all techniques, it is also important to consider the absolute difference in the time scales realized by each technique. A satellite can move several kilometers per second. So, if we require that the satellite orbits obtained from the different techniques should be consistent on the millimeter level, the offsets between the different time scales need to be known to better than 0.1–1 μs (depending on the orbit height). As already stated in Sect. 2.1, the offset in the VLBI time scale is probably less than 1 μs, thus it probably needs to be improved. For more discussion on this topic, see Anderson et al. (2018).

4 Summary

In this article we have reviewed the VLBI technique, discussed why the technique needs precise clocks, and assessed the precession needed. As indicated both by the discussion about coherence (Sect. 2.1) and the simulations (Sect. 2.2) hydrogen masers with an ASD of at least 10^{-14} @ 50 min are needed to guarantee the best performance. This is true for both the current VLBI system as well as for the future VGOS. Our results also indicate that since the atmosphere is the limiting factor, there is no point in using more precise clocks.

Acknowledgements We are very grateful to the reviewer of this article for providing many very detailed comments and suggestions beyond the normal level of a review. These helped us to significantly improve the quality of the paper.

References

Z. Altamimi, P. Rebischung, L. Métivier, X. Collilieux, ITRF2014: a new release of the international terrestrial reference frame modeling nonlinear station motions. J. Geophys. Res. **121**, 6109–6131 (2016). https://doi.org/10.1002/2016JB013098

J.M. Anderson, G. Beyerle, S. Glaser, L. Liu, B. Männel, T. Nilsson, R. Heinkelmann, H. Schuh, Simulations of VLBI observations of a geodetic satellite providing co-location in space. J. Geod. 2018, in press. https://doi.org/10.1007/s00190-018-1115-5

Y. Bar-Sever, B. Haines, W. Bertiger, S. Desai, S. Wu, Geodetic reference antenna in space (GRASP)—a mission to enhance space-based geodesy, in *Proc. 2nd Colloquium Scientific and Fundamental Aspects of the Galileo Programme*, Padova, Italy (2009)

R. Biancale, A. Pollet, D. Coulot, M. Mandea, E-GRASP/eratosthenes: a mission proposal for millimetric TRF realization, in *EGU General Assembly Conference Abstracts*, vol. 19 (2017), p. 8752

J.S. Border, Innovations in Delta Differential One-way Range: from Viking to Mars Science Laboratory, in *Proc. 21st Int. Sym. Space Flight Dynamics*, Toulouse, France, 28 Sept.–02 Oct., 2009, http://issfd.or/ISSFD_2009/OrbitDeterminationI/Border.pdf

D.A. Duev, G. Molera Calvés, S.V. Pogrebenko, L.I. Gurvits, G. Cimó, T. Bocanegra Bahamon, Spacecraft VLBI and Doppler tracking: algorithms and implementation. Astron. Astrophys. **541**, A43 (2012). https://doi.org/10.1051/0004-6361/201218885

D.A. Duev, S.V. Pogrebenko, G. Cimo, G. Molera Calvés, T.M. Bocanegra T.M. Bahamon, L.I. Gurvits, M.M. Kettenis, J. Kania, V. Tudose, P. Rosenblatt, J.-C. Marty, V. Lainey, P. de Vicente, J. Quick, M. Nickola, A. Neidhardt, G. Kronschnabl, C. Plötz, R. Haas, M. Lindqvist, A. Orlati, A.V. Ipatov, M.A. Kharinov, A.G. Mikhailov, J.E.J. Lovell, J.N. McCallum, J. Stevens, S.A. Gulyaev, T. Natush, S. Weston, W.H. Wang, B. Xia, W.J. Yang, L-F. Hao, J. Kallunki, O. Witasse, Planetary Radio Interferometry and Doppler Experiment (PRIDE) technique: a test case of the Mars express phobos y-by. Astron. Astrophys. (2016). https://doi.org/10.1051/0004-6361/201628869

E. Himwich, A. Bertarini, B. Corey, K. Baver, D. Gordon, L. La Porta, Impact of station clocks on UT1-TAI estimates, in *Proc. 23rd European VLBI Group for Geodesy and Astrometry Working Meeting*, ed. by R. Haas, G. Elgered (Gothenburg, Sweden, 2017), pp. 12–176

T. Nilsson, R. Haas, Impact of atmospheric turbulence on geodetic very long baseline interferometry. J. Geophys. Res. **115**(B03), 407 (2010). https://doi.org/10.1029/2009JB006579

T. Nilsson, B. Soja, M. Karbon, R. Heinkelmann, H. Schuh, Application of Kalman filtering in VLBI data analysis. Earth Planets Space **67**, 1–9 (2015)

A. Pany, J. Böhm, D. MacMillan, H. Schuh, T. Nilsson, J. Wresnik, Monte Carlo simulations of the impact of troposphere, clock and measurement errors on the repeatability of VLBI positions. J. Geod. **85**, 39–50 (2011). https://doi.org/10.1007/s00190-010-0415-1

B. Petrachenko, A. Niell, D. Behrend, B. Corey, J. Böhm, P. Charlot, A. Collioud, J. Gipson, R. Haas, T. Hobiger, Y. Koyama, D. MacMillan, T. Nilsson, A. Pany, G. Tuccari, A. Whitney, J. Wresnik, Design Aspects of the VLBI2010 System. Progress Report of the VLBI2010 Committee. NASA Technical Memorandum, NASA/TM-2009-214180, 58 pp. (June 2009)

W.T. Petrachenko, A.E. Niell, B.E. Corey, D. Behrend, H. Schuh, J. Wresnik, VLBI2010: next generation VLBI system for geodesy and astrometry, in *Geodesy for Planet Earth, International Association of Geodesy Symposia*, vol. 136, ed. by S. Kenyon, M.C. Pacino, U. Marti (Springer, Berlin, 2012), pp. 999–1006. ISBN 978-3-642-20337-4

L. Plank, VLBI satellite tracking for the realization of frame ties. PhD Thesis, Technische Universität Wien, Vienna, Austria (2013)

L. Plank, J. Böhm, H. Schuh, Precise station positions from VLBI observations to satellites: a simulation study. J. Geod. **88**(7), 659–673 (2014). https://doi.org/10.1007/s00190-014-0712-1

L. Plank, A. Hellerschmied, J. McCallum, J. Böhm, J. Lovell, VLBI observations of GNSS-satellites: from scheduling to analysis. J. Geod. **91**(7), 867–880 (2015). https://doi.org/10.1007/s00190-016-0992-8

C. Rieck, R. Haas, P. Jarlemark, K. Jaldehag, VLBI frequency transfer using CONT11, in *Proc. of the 26th European Frequency and Time Forum*, Gothenburg, Sweden (2012). http://publications.lib.chalmers.se/records/fulltext/171390/local_171390.pdf

A.E.E. Rogers, Very long baseline interferometry with large effective bandwidth for phase-delay measurements. Radio Sci. **5**(10), 1239–1247 (1970). https://doi.org/10.1029/RS005i010p01239

A.E.E. Rogers, J.M. Moran, Coherence limits for very-long- baseline interferometry. IEEE Trans. Instrum. Meas. **IM-30**(4), 283–286 (1981)

O.J. Sovers, J.L. Fanselow, C.S. Jacobs, Astrometry and geodesy with radio interferometry: experiments, models, results. Rev. Mod. Phys. **70**(4), 1393–1454 (1998)

J. Sun, J. Böhm, T. Nilsson, H. Krásná, S. Böhm, H. Schuh, New VLBI2010 scheduling strategies and implications on the terrestrial reference frames. J. Geod. **88**(5), 449–461 (2014). https://doi.org/10.1007/s00190-014-0697-9

G. Tang, J. Cao, S. Han, S. Hu, T. Ren, L. Chen, J. Sun, M. Wang, Y. Li, L. Li, Research on lunar radio measurements by Chang'E-3, in *IVS 2014 General Meeting Proceedings "VGOS: The New VLBI Network"*, ed. by D. Behrend, K.D. Baver, K.L. Armstrong (Science Press, Beijing, 2014), pp. 473–477. ftp://ivscc.gsfc.nasa.gov/pub/general-meeting/2014/pdf/101_Tang_etal.pdf

A.R. Thompson, J.M. Moran, G.W. Swenson, *Interferometry and Synthesis in Radio Astronomy*, 2nd edn. (Wiley, New York, 2007)

O. Titov, Estimation of the subdiurnal UT1-UTC variations by the least squares collocation method. Astron. Astrophys. Trans. **18**(6), 779–792 (2000). https://doi.org/10.1080/10556790008208172

V. Tornatore, R. Haas, S. Casey, D. Duev, S. Pogrebenko, G. Molera Calvés, Direct VLBI observations of global navigation satellite system signals, in *International Association of Geodesy Symposia*. Proc. IAG General Assembly, Melbourne, 2011, vol. 139 (2014), pp. 247–252. https://doi.org/10.1007/978-3-642-37222-3_32. ISBN/ISSN: 978-3-642-37221-6

A.R. Whitney, Precision Geodesy and Astrometry Via Very-Long-Baseline Interferometry. Ph.D. Thesis, MIT (1974)

Space Sci Rev (2017) 212:1433–1451
DOI 10.1007/s11214-017-0424-y

Survey of Capabilities and Applications of Accurate Clocks: Directions for Planetary Science

Véronique Dehant[1] **· Ryan Park**[2] **· Dominic Dirkx**[3] **·**
Luciano Iess[4] **· Gregory Neumann**[5] **· Slava Turyshev**[2] **·**
Tim Van Hoolst[1]

Received: 29 December 2016 / Accepted: 19 September 2017 / Published online: 16 October 2017

Abstract For planetary science, accurate clocks are mainly used as part of an onboard radioscience transponder. In the case of two-way radio data, the dominating data type for planetary radioscience, an accurate spacecraft clock is not necessary since the measurements can be calibrated using high-precision clocks on Earth. In the case of one-way radio data, however, an accurate clock can make the precision of one-way radio data be comparable to the two-way data, and possibly better since only one leg of radio path would be affected by the media. This article addresses several ways to improve observations for planetary science, either by improving the onboard clock or by using further variants of the classical radioscience methods, e.g., Same Beam Interferometry (SBI). For a clock to be useful for planetary science, we conclude that it must have at least a short-time stability (< 1,000 s) better than 10^{-13} and its size be substantially miniaturized. A special case of using laser ranging to the Moon and the implication of having an accurate clock is shown as an example.

Keywords Radioscience · Atomic clock · Positioning

High Performance Clocks with Special Emphasis on Geodesy and Geophysics and Applications to Other Bodies of the Solar System
Edited by Rafael Rodrigo, Véronique Dehant, Leonid Gurvits, Michael Kramer, Ryan Park, Peter Wolf and John Zarnecki

✉ V. Dehant
v.dehant@oma.be

1 Royal Observatory of Belgium, Uccle, Belgium

2 Jet Propulsion Laboratory/California Institute of Technology, Pasadena, USA

3 Delft University of Technology, Delft, The Netherlands

4 University of Rome, Rome, Italy

5 NASA Goddard Space Flight Center, Greenbelt, USA

Table 1 Ranges of different frequency bands used in radioscience

Frequency band	uplink	downlink
UHF	[390 MHz, 405 MHz]	[435 MHz, 450 MHz]
S-band	[2.026 GHz, 2.11 GHz]	[2.2 GHz, 2.29 GHz]
X-band	[7.145 GHz, 7.19 GHz]	[8.4 GHz, 8.45 GHz]
Ka-band	[34.2 GHz, 34.7 GHz]	[31.8 GHz, 32.3 GHz]

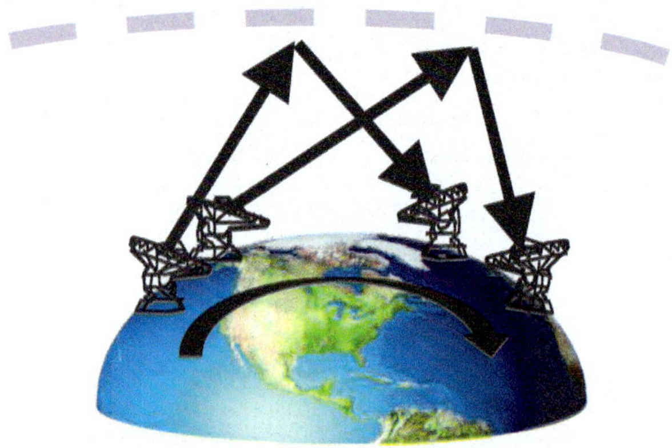

Fig. 1 Illustration of two-way radio measurements over a particular integration time

1 Survey of Capabilities of Accurate Clocks and Transponders in Space Science

Planetary science instruments/measurements using clocks or equivalent frequency standards are mainly for radioscience and laser instruments.

1.1 Radioscience

Radioscience instruments provide gravity, tide, and rotation measurements, from which the interior property of Solar System bodies can be deduced (see e.g. Folkner et al. 1997; Kuchynka et al. 2013, 2014; Le Maistre et al. 2012; Dehant et al. 2009, 2010, 2012a; Park et al. 2013, 2016). These observations are typically obtained using radio frequencies in UHF, S-band, X-band, or Ka-band. The frequency ranges for these radio bands as reserved for space radio links are presented in Table 1.

The measurements used to achieve the geodesy objectives are the Doppler shifts (i.e., for measuring line-of-sight velocity) and time of flight (i.e., for measuring distance) as measured at NASA Deep Space Network (DSN) stations or ESA ESTRACK tracking stations on Earth. These ground stations consist of 34 m/35 m and 70 m antennas. Similar 65 m antennas exist as well in Russia. Other equivalent stations (35 m and 64 m) are existing in China and Japan.

For two-way link between a spacecraft and an Earth's station (see Fig. 1), the measured Doppler shift can be written as (Moyer 2000):

$$\Delta f = \frac{f_T(t_1)}{cT_c}\left[\rho(t_2) - \rho(t_1)\right] \tag{1}$$

where f_T is the frequency transmitted at the spacecraft, c is the speed of light, and $\rho(t_1)$ and $\rho(t_2)$ are the spacecraft-station ranges at the start-time (t_1) and end-time (t_2) of the

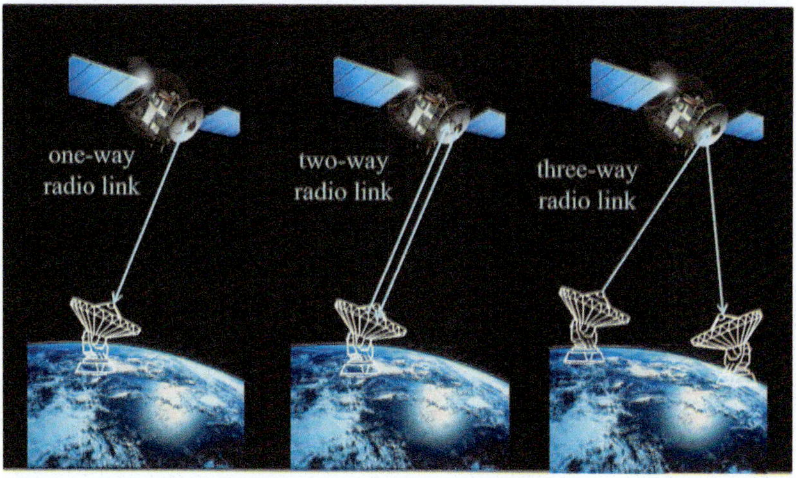

Fig. 2 Representation of the one-way, two-way and three-way radio-links

Doppler count time T_c, respectively. The ranging measurements consist of time differences between the signal sent to the spacecraft and received at the ground station. The geometry, or equivalently the orientation of the line-of-sight direction, changes over the sample time. From these measurements, one can reconstruct the motion of a spacecraft around a planet or moon (or on its way to the target celestial object) or of a lander or a rover on a planet or moon surface with respect to the Earth. One can as well measure the relative motion of a lander or a rover with respect to a spacecraft. The radio signals are often used as a "two-way" signal (see Figs. 1 and 2). For a two-way measurement, the reference frequency is provided on an Earth station, the signal is transmitted from the station, received at a spacecraft, and transmitted back coherently to the same Earth station. In this configuration, the same Earth station is used for transmission and reception of the signal. The "three-way" configuration is sometimes used, in which case, the receiving station is different from the transmitted station (see Fig. 2). The "three-way" data needs to be corrected for an offset between the clocks at the two stations, which is a well-understood practice and its accuracy is comparable to two-way data. Obviously, in these configurations, the reference clocks are calibrated on Earth and an onboard clock is not needed as long as the data is acquired coherently. The ESTRACK or DSN ground stations use hydrogen maser as their references (at the level of 10^{-15} stability). However, for the transponders to be coherent (retransmitting at a fixed ratio with respect to the reference frequency with no phase lag induced from the instrument), they are usually based on phase-lock loops and involve an internal oscillator. This oscillator must be as stable as possible but practically a quartz is sufficiently stable.

On the other hand, one could equivalently use an Ultra Stable Oscillator (USO) onboard the spacecraft and receive on Earth, or vice-versa; this is the so-called "one-way" radio tracking (see Fig. 2). The precision of the one-way measurements is mainly limited by the stability of an onboard USO, usually at the intrinsic stability level of 10^{-13}. In practice, two-way Doppler and ranging are mainly used for radioscience applications since present-day USOs are very sensitive to thermal environment and it is hard to achieve the intrinsic instrument stability (Asmar et al. 2005). In addition, a spacecraft must have a transponder for communication purposes, thus there always exists a two-way capability.

Presently the radioscience observation allows us to provide information on the global orbit of the spacecraft and its changes in space and time (precise orbit positioning), on gravitational and non-gravitational perturbations on the spacecraft trajectory and their time variations, including tidal or seasonal atmospheric effects, on the acceleration undergone by the spacecraft at a particular space and time (gravity above particular targets), on the orientation and rotation parameters of the planet or moon around which the spacecraft is in orbit or flies by. Two-way Doppler and ranging are used at present. If precise clocks are available on the spacecraft, one-way measurements could supplement these measurements. A benefit for one-way measurements in planetary missions can also be found when using it for atmospheric occultation, which can be used to determine atmospheric properties of a target body.

1.2 Interplanetary Laser Ranging

A laser altimeter enables distance measurements from a spacecraft to the surface of the orbiting body. The distance is determined by precisely measuring the round-trip light time of a laser pulse from the instrument to the surface and back to the instrument. The most important limitations of these measurements are the spacecraft position and attitude knowledge, and the so-called range walk due to signal strength variations and the spread of the laser pulse waveform by surface variations within the laser beam footprint.

A laser altimeter can also be used for measuring distance between Earth and a spacecraft or between Earth and a corner cube on a spacecraft or on the surface of a planetary object. The precision of this measurement is related to the frequency stability of the onboard clock or ground station USO.

In Satellite Laser Ranging (SLR), arrays of retroreflector corner cubes on spacecraft serve as transponders (Degnan 1993), providing sub-cm accuracy and precision range measurements. With return signal declining as the fourth power of distance, the limitations of such passive transponders beyond Earth orbit are clearly apparent. Active interplanetary transponder experiments using laser altimeters in cruise were first proposed in 1992 after the September launch of the Mars Observer spacecraft, carrying the Mars Orbiter Laser Altimeter (MOLA) as part of its instrument suite. The goal was to test the MOLA laser by firing at Earth and attempting to detect its signal, and to fire a ground laser at MOLA. The request never received serious consideration and the spacecraft was lost before attaining Mars orbit. The Mars Global Surveyor project did approve such an experiment with the reflight of MOLA in 1997, but it was canceled due to a spacecraft anomaly. An attempt by the NEAR-Shoemaker spacecraft to range to Earth prior to a flyby in 1999 was unsuccessful owing to poor visibility over several days at the one ground station equipped to receive 1064 nm (very near infrared) signals. Eventually, in 2005, a one-way link was achieved to the Mars Global Surveyor (MGS) spacecraft at a distance of 80 million km from Earth, with a precision of 2 milliseconds. Improved precision could have been achieved if the clock element of MOLA had not already failed (Abshire et al. 2006).

Two-way interplanetary ranging was eventually demonstrated successfully in 2005 (Smith et al. 2006) with the Mercury Laser Altimeter (MLA) on board the MESSENGER spacecraft. The ranging precision was limited somewhat by the ~ 6 ns MLA pulse width and the weak signal received by the instrument on the spacecraft, just at the detection limit. This two-way asynchronous transponder experiment (Degnan 2002) achieved a few dozen downlink and uplink data that corresponded to the receive times predicted for a range of nearly 24 million kilometers. Solutions for the range and range rate as well as clock offset were obtained with a precision of 20 cm, 15 cm/s, and 0.7 ns respectively from the 104 data points acquired. The experiment succeeded in part due to the precision of the onboard USO that provided timing signals and was also monitored from the ground, but by the time the experiment could be repeated four days later, the clock had drifted by several microseconds so that a longer-baseline solution could not be obtained.

Building on this success, the Lunar Reconnaissance Orbiter (LRO) incorporated a one-way Laser Ranging (LR) experiment (Zuber et al. 2010), providing operational laser tracking for the first time of a satellite beyond Earth orbit. The Lunar Orbiter Laser Altimeter (LOLA) instrument had a ranging precision of 10 cm. To attain the stringent 50-m total position orbit requirement, a receiver telescope was articulated with the LRO high gain antenna (HGA) and coupled to the LOLA timing system to allow an Earth laser station to range to LRO in a one-way transponder mode whenever in view. LRO-LR operation was conducted for a total of $> 4{,}000$ hours by 11 ground stations in 5 years, from 30 June 2009 to 30 September 2014. LR data achieves a 5–10 cm precision after conversion into 5 s normal points (Mao et al. 2012, 2017). The spacecraft clock is a commercial ovenized crystal oscillator (Symmetricom 9500B) similar to the USOs flown routinely on other spacecraft with a short-term stability of $< 2 \times 10^{-13}$ Allan deviation (Bauer et al. 2016, 2017). For the two-hour duration of a lunar orbit, random walk errors are $\sim$ 1–2 ns. The one-way experiment relied upon the precision of both the spacecraft clock and multiple instances of ground station clocks, often with unpredictable offsets that introduced many additional parameters into the orbit estimation.

The GGAO (Goddard Geophysical and Astronomical Observatory) ground station relied at first on a GPS-steered clock for observation epoch time; this proved inadequate and was eventually replaced by a co-located H-maser. Other stations used masers during part of the experiment, or relied on Cs and Rb atomic clocks with variable stabilities and drift rates.

The LR observables for orbit determination are the matched transmit and receive times, corrected for system and atmospheric delay, formed into 5 s centered averages or "normal points". Modeling the observable time difference requires a fully relativistic treatment of both clocks as well as the clock offsets and rates. Additional care to model the spacecraft clock required estimation of its intrinsic drift and aging, as well as its residual thermal response and its unpredictable or random walk noise. Several teams, using different software systems have utilized portions of the LR data to study orbit determination: the JPL navigation software package MONTE (Buccino et al. 2016), the software developed jointly at Delft and DLR using the Tudat software (Bauer et al. 2016, 2017), and the GEODYN2 software at GSFC (Mao et al. 2017). All three efforts obtained tracking residuals in the range of 10–30 cm, and found that performance on a par with 2-way radiometric tracking could be obtained over relatively short (1–2 day) arcs, for which the random walk of the USO was not severe. Some results have been obtained with longer (5–7 day) arcs, but the number of extra parameters to model the clock thus introduced limits the strength of one-way ranging. In combination with radiometric ranging and Doppler tracking, the incorporation of laser ranging data improved the LRO orbits, with the improvement correlated with the amount of tracking data and modeling sophistication.

The Moon is, however, a special example because the most important factor in orbit determination has been the use of high-degree gravity potential models derived from the GRAIL mission (Mazarico et al. 2013), and the best results obtained for LR likewise rely on the best resolution of gravity, particularly on the far-side where tracking from Earth is unavailable.

For well-determined orbits, the LR simultaneous data passes (up to four stations at a time) can be used to affect a time transfer between stations within the field of view of the LR telescope (about 12,000 km). The GGAO maser and the nearby US Naval Observatory master clocks can be correlated at well within 1 ns using data collected from GPS receivers, given that ionospheric sources of error are common to both stations and therefore largely cancel, resulting in residual differences over 20 months of < 10 ns after removing long-wavelength trends. Thus, strong bounds may be placed on ground station clock errors; the main issues reside in the behavior of the spacecraft clock.

Due to the very similar required systems, there is great synergy between laser ranging and laser communications (Hemmati et al. 2009). The LADEE (Lunar Atmosphere and Dust Environment Explorer) spacecraft was the first to apply such a communications system beyond Earth orbit (Elphic and Russell 2015), using the Lunar Laser Communication Demonstration (LLCD) (Boroson et al. 2014). From the communications data, range data with a precision at the several cm level was obtained.

1.3 Accuracies/Stabilities of Clocks/USOs Currently Available for Space Science

As mentioned already, no clock is used onboard spacecraft for planetary science except for USOs with 10^{-13} stability used for one-way links. To reach this stability, the onboard USO must be thermally controlled. To make use of the Doppler effect for navigation and radioscience in a one-way link, it is necessary to know the frequency of the radio signal transmitted from the spacecraft at that level. Then the received frequency can be analyzed to calculate the line-of-sight speed of the spacecraft. Many interplanetary spacecraft carry a USO for this purpose, ensuring thermal stabilization to keep its output frequency stable. The long-term stability of a USO might be a problem if not calibrated against the stability of an Earth-based maser, which can be done using a model for the frequency changes over time.

The present-day precision of the radioscience measurements is limited by the noise arising on the signal from other sources, such as the errors due to solar plasma, ionosphere, troposphere, and non-gravitational forces such as maneuvers and angular momentum desaturations.

The precision of planetary laser ranging is limited by the pulse width, which is typically 10–100 ps (single-shot precision of 3–30 mm). The influence of clock noise for two-way ranging is limited, as discussed in Sect. 3.3. Measurement processing accuracy is limited to the several mm level by uncertainties in tropospheric corrections, ground station position uncertainties, and uncalibrated system instabilities.

2 Survey of Applications of Accurate Clocks and Transponders in Planetary Science

2.1 Global Static Gravitational Field Determined from a Spacecraft

When a spacecraft is orbiting a planet, it is perturbed by the gravity field of the central body. If the planet is homogeneous and spherical, the gravitational field is equivalent to that of a point mass, and the orbital motion of the spacecraft due to this gravity field would follow a Keplerian motion. However, planets, such as Earth, Mars, Venus, or Mercury, are not spherical and their masses inside are not uniformly distributed. For an oblate planet, one of the most important variations in the orbit corresponds to the precession in the longitude of the node of the orbit and is proportional to J_2 (the form factor of the planet), i.e. the coefficient of the first even zonal harmonics of the external gravitational potential. It provides information on the mass distribution inside the planet since J_2 is proportional to the difference between the polar moment of inertia (C) and the equatorial moment of inertia (A), representing the dynamical flattening of the planet. For a rapidly rotating planet like Mars or the

Earth, the contribution of the higher degree harmonics to the orbital drift is significantly lower than the J_2 contribution, because of the ellipsoidal distribution of the gravity field. *The limitations here are not arising from the clock but well from the other forces entering into the determination of the gravitational field, i.e., effects of angular momentum desaturation, maneuvers, atmospheric drag, solar radiation pressure effects on the spacecraft, as well as solar plasma and Earth ionospheric/tropospheric effects on the signal.* However, as mentioned previously, if a clock is available onboard the spacecraft, it is possible to add one-way communication between the Earth and the spacecraft, allowing to use passive telescopes on the surface of the Earth. The one-way links add data, which is important for a global computation of the gravity field. Moreover, assuming the clock noise could be reduced to provide comparable data noise as a two-way link, the one-way link would provide about a factor of $\sim \sqrt{2}$ improvement since the error due to media are cut in half.

2.2 Time Variation of the Gravitational Field as Determined from a Spacecraft

For a planet with an atmosphere, changes in the mass distribution and wind effects in the atmosphere may induce changes in the solid planet angular momentum and changes in the gravity coefficients. For Mars, about one fourth of the atmosphere of Mars is participating in the sublimation and condensation process of the CO_2 in the atmosphere and ice caps. The CO_2 seasonal cycle of Mars atmosphere induces mass exchange between the pole caps and the atmosphere, which in turn induces seasonal variations of the gravitational potential as well as of the rotation rate of the planet. The expected contributions to the gravity zonal harmonics variations of degree 2 to 5 are on the order of a few hundredths of percent of their static values as observed and expected from Global Circulation Model (GCM) of Mars' atmosphere. In turn, this small amplitude of the zonal harmonics variations generates small orbital perturbations, which needs an accurate reconstruction of the orbiter motion to be detected. *Again, the limitations are not arising from the clocks but from the non-gravitational corrections of the spacecraft acceleration.*

2.3 Time Variation of the Gravitational Field Induced by the Tides as Determined from a Spacecraft

Tides are interesting phenomena as they can be used for the determination of core properties such as the core density and size. The tides depend on the interior structure, e.g. there are larger tidal displacements for a body with a fluid core or a global subsurface ocean. If the tides are observed, it may be possible to determine the interior structure since the driving forces are usually known precisely. For Venus (Konopliv and Yoder 1996) and Mars (e.g. Konopliv et al. 2011), the tidal k_2 Love numbers have been observed and the physical properties of the core have been determined. At first, the core density/dimension can be determined, which are as well a tread off with respect to the mantle density profile. Equivalently, the tides of the icy moons are as well interesting to observe as they reveal the existence or not of a subsurface ocean. In this case as well, the ocean density can be determined in a tread-off approach with respect to the ice rigidity and the crustal thickness (e.g. see Baland et al. 2014; Mitri et al. 2014, for Titan; see Wu et al. 2001; Wahr et al. 2006, for Europa; and Moore and Schubert 2003, for Ganymede and Callisto). *Again, here the limitations are not arising from the clocks but from the non-understood non-gravitational corrections of the spacecraft acceleration.*

2.4 Time Variation of a Lander Position for Rotation and Tide Determination

The observables are Doppler and ranging in X-band or Ka-band or in both using the radio links between landers at the surface of Mars and the Earth surface tracking stations. From these observables, it is possible to determine Mars' rotation and orientation in space as well as the lander position as a function of time. These observations are further used to determine properties of Mars (or equivalently the Moon) interior, such as the core density, the fraction of core mass due to the presence of a light element (sulfur with the remaining mass assumed to be iron), the radius of the core etc. Similarly, rotational changes, such as changes in the length-of-day of Mars, may be determined by tracking the motion of a lander position in space, which enable to provide information on the atmosphere of Mars.

In addition to the determination of precession and nutation for Mars (i.e. variations of the orientation of Mars in space due to gravitational torque that the Sun exerts on the planet), the technique can be used for the determination of physical libration, (i.e. physical oscillatory response of a moon or planet in spin-orbit resonance to the gravitational torque exerted by the mother planet or the Sun, respectively) if one would have a lander on the surface of Mercury for instance, or on the surface of on icy moon of Jupiter.

Tidal effects do not only affect the dynamics of a spacecraft but as well of a lander. Periodic time variations of the lander position with respect to the planet or moon are induced either by the tides or by seasonal loading effects, which can be determined thus from radioscience with lander.

Here, again, the limitations are not arising from the clocks but from the non-gravitational corrections of the measurements.

3 Future Applications

3.1 Clocks on Spacecraft and Applications

Before we can really benefit from high-precision onboard clock, further advancements must be first made on the modeling and determination of non-gravitational forces acting on a satellite. These non-gravitational forces could be evaluated with the help of precise onboard accelerometers. In the case that all the forces acting on the satellite are well understood, with the exception of the gravitational forces (both static and time-varying), high-precision clock can be of interest.

Having an accurate clock onboard will allow one-way link with improved noise (i.e., a factor of $\sim\sqrt{2}$ since only one leg of radio path would be affected by the media error) than the two-way link. A gain of one order of magnitude on the onboard clock stability (i.e., 10^{-14}) would immediately translate into improvements in Doppler and ranging precision. However, since the Doppler data noise are limited by the media effects (e.g., solar plasma, ionospheric and tropospheric errors), use of a better clock is only meaningful if these errors can also be improved.

Combining several frequencies would help decreasing the Doppler data noise, similar to what was done on GPS and Galileo. The use of at least two frequencies (e.g. X-band and Ka-band) is also a potential approach for calibrating the errors due to solar plasma and Earth ionosphere. However, we will also need to calibrate the tropospheric effect at the comparable level. This applies to both the Earth atmosphere as well as the planet atmosphere if the spacecraft signal passes through the planet atmosphere. The Earth tropospheric effect can be calibrated using complementary data such as Advanced Water Vapor Radiometer (AWVR) and/or GNSS.

Thus, one way of taking the advantage of accurate clocks in space is using it for one-way radio link, which could increase the number of measurements and provide better geometric coverage. Also, high-precision timing, together with the usage of several frequencies and AWVR, would allow improving the precision of the Doppler and ranging measurements.

3.2 Same Beam Interferometry (SBI)

Recently a new method has been proposed for gaining an order of magnitude precision improvement in the determination of differenced one-way or two-way measurement, i.e., Same Beam Interferometry (SBI). SBI entails the simultaneous tracking of two or more landers from a single ground station in a coherent, two-way mode (Bender 1994; Fermi et al. 2008; Iess et al. 2008; Gregnanin et al. 2012, 2014). The signals received back from the landers are recorded and combined in an interferometric mode to form a differential phase proportional to the difference in distance between the landers and the ground station. The interference fringes retain information only about rotation and tides, while common noise sources such as interplanetary plasma, the Earth troposphere and ionosphere cancel out. The relative motion of the Earth and Mars is also rejected when the differential phase is formed. The noise cancellation is limited only by the different path delays in the Martian atmosphere and ionosphere (to a much lesser extent), and the small gradients of the interplanetary plasma density across the two line-of-sights.

The availability of a network of landers on a planet, a moon, or even an asteroid would enable this new and powerful method for estimating the target's tidal and rotational states. The availability of two landers simultaneously viewed from Earth (from a single ground station) allows their simultaneous tracking in a coherent two-way mode as explained above. The crucial requirement of the interferometer is an excellent coherence between the transponders. Thanks to the common mode rejection, the main sources of measurement errors affecting single link (mostly propagation noises and clock drifts) are effectively suppressed in a differential (interferometric) configuration. This method requires the stability of the Earth ground station clocks and again do not need further improvement in clock in space.

Application of this technique has been foreseen/proposed in the frame of a network on the Moon and on Mars. In the present framework, we consider direct-to-Earth links using transponders on the landers for radioscience (and partly telecommunication) and examine different options for future configurations. All links need to be two-way coherent, with the signal transmitted from a single ground station. No ultra-stable oscillator (USO) is necessary on the landers. Indeed, one-way links would destroy the phase coherency required by SBI, and would seriously degrade even the performances of conventional Doppler and range measurements. This might be relaxed if very precise clocks would be available. However, this would add unnecessary weight on the payload, if one considers the present-day weight of atomic clocks for space. If miniaturization of the atomic clock were available, radioscience would benefit from it. This advance would go in pair with using several frequencies in space, as explained before, a little bit like the improvement of Galileo for positioning. In addition, all the applications foreseen with Galileo could be envisaged with the new system. For the Galileo satellites, the onboard passive hydrogen maser clocks' precision is four times better than the onboard rubidium atomic clocks of the previous GPS generation and is estimated at 1 second per 3 million years (a timing error of a nanosecond per day), which translates into a sub-millimeter positional error on Earth's surface. For such a probe and clock around a planet, the error on the distance would be at the level of ~ 10 centimeters.

Fig. 3 Radio links for geodesy. "Up" means that the frequency is the one sent from Earth to space (which is for the X-band for instance near 7.15 GHz), and "down" means that the frequency is the one sent from space to Earth (which is for the X-band for instance near 8.4 GHz)

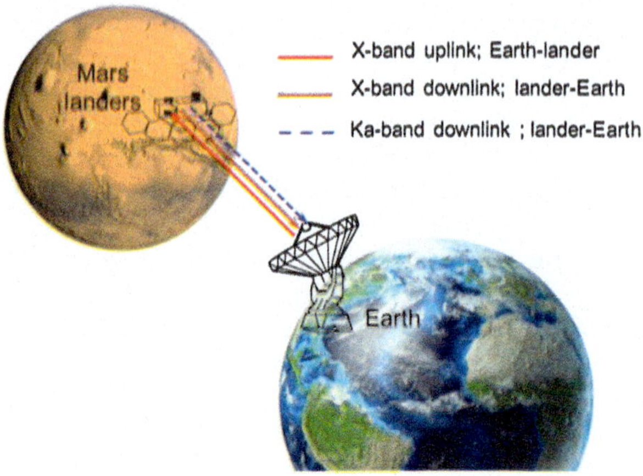

This is the level achievable by present-day two-way Ka-band Doppler. However, the position determined from Doppler measurements benefiting from such clock precision would be at the sub-centimeter level.

Several radio link configurations may be envisaged:

In the first option [Option 1], the links are enabled by conventional X-band transponder such as Mars Express, Venus Express, Rosetta, BepiColombo (without the Ka-band channel), and the LaRa (Lander Radioscience) onboard the surface platform of ExoMars 2020 or RISE (Rotation and Interior Structure Experiment) on the 2018 InSIGHT (Interior exploration using Seismic Investigations, Geodesy, and Heat Transport). This is the simplest instrumentation, but not the top performer, while it would already help the present-day precision of the measurements.

In Option 2, the transponder enables a dual frequency downlink at X- and Ka-band (see Fig. 3 for a lander-Earth case), both coherent with the X-band uplink. This configuration enables the removal of dispersive path delays due to charged particles in the downlink. The extent to which the downlink information can be used to assess the uplink contributions depends on the correlations between the two terms and must be safely assumed as unknown. Option 2 would therefore reduce the plasma noise by a factor 1.4 (approx.).

In Option 3, the transponder enables two separate two-way Ka-band links. This solution appears much satisfactory in terms of plasma noise level and very simple, but Ka-band transponders are technically much more complicated by the necessity of using waveguide in the transponder. The power consumption is also higher than an X-band transponder. One may consider the selection of the uplink frequency according to operational requirements and the availability of other nodes provided by different agencies.

It must be noted that Option 2 offers limited benefits over a pure X-band radio link, both for standard radiometric measurements (Doppler and range) and SBI. Option 3 offers a very effective suppression of plasma noise under most operational conditions (except for Sun-Earth-probe angles smaller than 15 degrees). Plasma noise is the dominant error source in X-band tracking systems.

A number of variants of these three basic options can be considered. The most important one entails the adoption of a Spread Spectrum (SS) signal structure. In this configuration the carrier, telemetry and telecommand are identified by a unique code, with a signal effectively occupying the entire allocated bandwidth. The carrier phase (the main observable in SBI and radiometric measurements) is reconstructed from the code. In 3G and 4G terrestrial

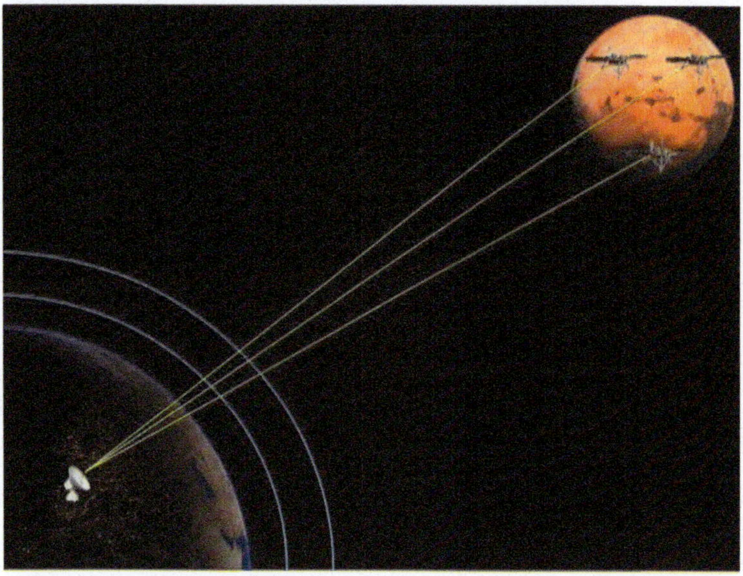

Fig. 4 The configuration of the Mars microwave interferometer. Two or three landers are simultaneously tracked from the same ground antenna. All signals use the same frequency, with the carrier phase reconstructed through a unique code for each lander. The differential phase retains only the effect of rotation and tides, while all noise contributions common to the links are cancelled out

cellular networks, the scheme is also known as Code Division Multiple Access (CDMA). SS communications are widely used in military space communications. European industry has a vast experience in this field.

For a network of landers, there are both operational and scientific motivations for adopting such an SBI scheme. Operationally, the entire network can be tracked and controlled from a single ground antenna, with considerable saving of antenna time. Orbiters can also be tracked simultaneously with landers. Science benefits are important for SBI as this configuration guarantees the maximum commonality between the radio links, as it will be explained later.

In summary, we have three options of SBI corresponding to the different options presented above: (1) X-band up-link from Earth to lander and X-band downlinks from lander to Earth; (2) X-band up-link from Earth to lander and X- and Ka-bands downlinks from lander to Earth; and (3) Ka-band up-link from Earth to lander and Ka-band downlinks from lander to Earth. Only the second option will allow studying the ionosphere of Mars in addition to the objectives of studying the global atmosphere and the interior of Mars from geodesy.

Thanks to the common mode rejection, the main sources of measurement errors affecting single links (mostly propagation noises and clock drifts) are effectively suppressed in a differential (interferometric) configuration. Interferometric observables can therefore provide an excellent determination of Mars' rotational parameters and tides. The concept of a microwave interferometer entails two or more transponders on Mars, simultaneously illuminated by a single radio antenna on the ground. Each transponder coherently retransmits a unique signal to ground, where each phase is measured and the phase differences are formed (see Fig. 4).

The measurement goal is to reach an accuracy of 1 mm (corresponding to 1/10 of the Ka-band wavelength) in each differential range, with integration times of 100–600 seconds.

The daily observation time can be limited to 1 hour in order to attain a suitable data set for geophysical interpretation. If the nodes are, say, 1,000 km apart, a 1 mm accuracy in the differential range corresponds to an accuracy of about 10^{-9} radians in the orientation of Mars. Differently from the case of a lunar interferometer, Mars' fast rotation allows an easy separation between tides and orientation changes with the use of just two landers. A particularly favorable positioning on the surface entails one transponder on the equator and another one shifted in latitude and longitude by 60 degrees. Other configurations are also possible and an efficient distribution can be studied.

3.3 Future Applications of Laser Ranging

Using experience with past transponder laser ranging experiments (see Sect. 1.2), in particular the one-way laser ranging data obtained operationally with the LRO spacecraft, laser ranging capabilities could be expanded to interplanetary distance (Degnan 2002). The very small (19 mm diameter) telescope aperture used on LRO was dictated by accommodation on an existing gimbaled antenna, whereas the LADEE spacecraft carried a purpose-built optical communication instrument with ~ 100 mm aperture and dedicated pointing. Neither system required more than 1 Watt of output laser power. Obviously, the technological development of a deep space system would require a larger power-aperture product. The LLCD space terminal required only modest clock stability ($\sim 8 \times 10^{-11}$) over the 2.5 s roundtrip time to meet its 200 ps requirement. Laser ranging in planetary missions would complement the tracking data with highly accurate range measurements (on the order of mm). Such tracking data would be especially attractive for the determination of solar system ephemerides, rotational variations, relativistic effects and other long-periodic effects that are encoded in the tracking data (Turyshev et al. 2010; Oberst et al. 2012; Dirkx 2015). The extraction of shorter-period effects, such as gravity field characteristics, will, in most cases, be more accurate using existing (two-way) radio Doppler techniques.

As with radio tracking, laser ranging can be performed in a one- or two-way mode (see Sect. 1.2). In the two-way mode, the contribution of clock noise to the error budget of the measurements is limited. An integrated timing error of 1–10 ps over 1–100 s typically below the noise level of two-way transponder laser ranging (Dirkx et al. 2015), is leading to a clock stability requirement of 10^{-10}–10^{-12} (at 1 s) and 10^{-12}–10^{-14} (at 100 s). Such requirements are not trivial, but are well within the capabilities of state-of-the-art systems. For one-way laser ranging, however, clock noise is the dominant source of measurement uncertainty, and poses a significant challenge during orbit determination, as discussed above for the case of LRO and shown in numerical simulations by Dirkx et al. (2015).

Using highly accurate space-based clocks would significantly reduce the number of clock parameters that need to be estimated during orbit determination from one-way data, thereby allowing these data to be used for the retrieval of a broader class of parameters. Since the science case for planetary laser ranging is especially strong for retrieving long-periodic signals, the system's long-period stability is crucial. For the case of one-way laser ranging to LRO (Buccino et al. 2016; Bauer et al. 2016, 2017; Mao et al. 2017), clock noise over longer integration times makes it difficult to achieve such stability, indicating that improvements in clock stability is highly desirable for future one-way systems. Additionally, it is shown by Dirkx et al. (2016) that the proper-to-coordinate time conversion must be updated (iteratively or concurrently) during the estimation process to fully utilize future laser ranging data sets, due to the large discrepancy between a priori state uncertainty and measurement noise level.

However, with two-way, two-frequency radio range measurements promising an attainable accuracy of 20 cm (Iess et al. 2014), it will be difficult for a one-way laser ranging system to be competitive. Dirkx (2015) have performed a comparative analysis of Doppler and range measurements for the retrieval of periodic signals from the tracking data, and conclude that a 20 cm range measurement becomes competitive with Doppler measurements (0.002 mm/s quality at 1,000 s integration time) only for signals with periods > 200 hours. For one-way laser ranging systems, outperforming radio range measurements would require 10^{-15} fractional stability for signals with a characteristics period of 200 hours and 2.10^{-17} stability for 1 year periods.

Although this could point to the need for highly accurate space-based clocks in planetary missions, the attainable range accuracy can (at present) be reached with a much smaller hardware suite using a two-way laser ranging system (which will also be about 2 orders of magnitude more accurate). Nevertheless, future advances in miniaturizing highly accurate clocks would allow one-way systems to become competitive with existing tracking systems. An additional application for one-way systems is in missions where the Earth-pointing requirement of the space-based laser system is impossible to meet, or where stray light would be excessive (as for missions to the outer solar system).

3.4 New JPL Facility to Explore Deep Lunar Interior

LLR presently is conducted using pulsed lasers with very high peak power (~ GW) and sub-nanosecond pulses. However, despite the very high peak power, the average power is quite modest (2 W). For instance, APOLLO uses a laser with 150 mJ pulses at a rate of 20 Hz, multi-photon detection (3–5 photons per pulse), and a good astronomy site to get to ~ 8 mm lunar range precision in single shot (Murphy et al. 2008). The primary limiting factor to the LLR accuracy today is the Earth's atmosphere, which imposes a lower limit of 5 mm for the current LLR operations. In addition, the accuracy of LLR is limited by the CCR (Corner Cube Retro-reflector) arrays currently on the Moon and their distribution on the lunar surface (Williams et al. 2009), and by the LLR model (Williams et al. 2004).

To further advance lunar research, JPL is building a new LLR facility at the Table Mountain Observatory (TMO) in California, with the objective of bringing major advances in the precision of LLR measurements. Equipped with a high-power (~ 1.1 kW) continuous wave (CW) laser, TMO will not only receive a much large flux of returned photons (~ 103 photons/sec is expected) and thereby obtain a concomitant higher signal-to-noise ratio than other facilities, it will also enter a new photon-rich regime in which differential LLR is feasible.

In a photon-rich regime, such as what we propose to enable for TMO, simultaneous ranging to two or three arrays is possible. With high photon flux (as above mentioned), the fundamental limitation to LLR accuracy is then no longer $\sqrt{N}$ where N is the number of photons, but the lack in knowledge of the atmospheric delay. The delay from Earth's atmosphere (~ 2.3 m at zenith) produces an uncertainty for single CCR ranging of ~ 8 mm after correction using the temperature, pressure, and humidity data. However, the differenced delay error from one telescope to two or three sets of lunar CCRs can be as little as ~ 30 μm. Also, in this regime, the station-specific errors are common mode. This will allow us to measure lunar librations with an accuracy of ~ 10 microarcsecond as referred to the lunar rotation angle. The new LLR facility will dramatically improve our knowledge of the lunar interior, beyond the contributions from the GRAIL mission and current LLR data (Williams et al. 2014).

Differencing ranges to different CCRs would reduce substantially atmospheric perturbations, one of the dominant terms in the current LLR error budget. This can be done by switching between two (or several) lunar targets. To demonstrate this, we estimate the LLR differential delay error due to the atmosphere relying on the extensive data collected by adaptive optics systems over the past few decades. The isoplanatic angle is defined as the angular separation over which the wavefront from two stars will differ by 1 radian RMS. The isoplanatic angle in the visible (0.6 μm) is $\sim$ 10 μrad (2 arcsecond); thus, two objects separated by this angle will have atmospheric delay contributions that differ on average by $\sim$ 100 nm. At 0.5 degree separation (the angular width of the Moon), this differential path error grows to $\sim$ 31 μm (using the Kolmogorov turbulence scaling law). This 31 μm error should decrease as $\sqrt{T/t_0}$, where t_0 is time it takes the wind to move the air across the two beams at the top of the troposphere ($\sim$ 5–10 sec). For an infrared laser (say, at 2.2 μm), the differential path difference over the lunar width would be $\sim$ 30.8 μm.

At any given time, the new facility at TMO will monitor the return photon fluxes between OCTL (Optical Communications Telescope Laboratory) and a single lunar CCR array. Unlike classic LLR, which measures the round-trip light-time from Earth to the Moon and back, the advanced LLR will track the phase of the returned amplitude-modulated microwave signal to determine the lunar range. At TMO, we plan to difference any of the two consecutive range determinations to form a new LLR observable, differenced lunar range, previously not available. The advantage of the new scheme is in the fact that it allows reaching the differenced range precision at 30 μm, which is the Earth atmospheric limit. The time-multiplexing between two sets of CCR's results in a factor of $\sqrt{2}$ increase in error; also, subtracting two ranges causes another factor of $\sqrt{2}$ increase in error for the differenced range measurement. Thus, the differenced range photon noise is $\sim$ 4 mm in 1 sec, reducing to 460 μm in 1 minute. Integrating this signal for 2 hours would get us down to 42 μm. The period of the libration signal is much longer than 2 hours, allowing for a precise determination of the relevant science parameters via the differenced LLR technique. With such a performance, we can reach quantitative goals even in the worst-case scenario of full Moon and small CCR arrays.

The combination of the new CCR on the Moon (Turyshev et al. 2013) and the new JPL facility would dramatically enhance our knowledge of the lunar interior, beyond the contributions from the GRAIL mission and current LLR measurements. The new differenced range data accurate to 30 μm-level would allow for an expansive study of the deep lunar interior including the core-mantle boundary shape, core rotation, and boundary-layer turbulence, inner solid core detection, deep mantle tidal rigidity and dissipation in a region of suspected partial melting, asymmetrical tidal response (Zhong et al. 2012), rheology-caused frequency-dependent tidal Love numbers (Efroimsky 2012; Nimmo et al. 2012; Williams and Boggs 2015), and stimulation of free librations (Rambaux and Williams 2011). Differenced LLR measurements would build on and enhance the results from the GRAIL mission (e.g., deep interior: study of the fluid core, detection of inner solid core, etc.).

3.5 The Special Case of the Moon

The Moon and instrumentation on its surface constitutes a special case as colocation of high-precision techniques and clocks would enable better reference frame realization. In the Apollo era, the use of passive laser reflectors on the lunar surface has been the most efficient measurement technique. However, the laser link margin is very small, permitting only very few ground stations to obtain valid measurements. Therefore, the next generation of the Lunar Laser Ranging (LLR) technique should aim for substantial improvement. Dehant et al. (2012b) proposed a one-way ranging concept involving laser transponders, which involves the use of a very stable frequency. This allows many more Satellite Laser Ranging (SLR) stations from the existing network on Earth to obtain measurements from the Moon. In addition, Dehant et al. (2012b) proposed to deploy and operate a microwave receiver/transmitter with precisely known mechanical local ties to the laser beacon/receiver, which will permit observations of the tangential position of the Moon with respect to the celestial frame. The ultimate objectives are threefold, the improvement of the reference frames for the Earth, a better understanding of the Moon's interior, and a better determination of the parameters of General Relativity, as described in that paper.

The interior objective is reached through the observation of the Moon libration, of which the amplitude is sensitive to the liquid core. A fluid core for instance does not necessarily share the rotation axis of the solid mantle because the oblateness of the core-mantle boundary is expected to be small and the viscosity has a negligible effect. While the equator of the solid Moon is subjected to a precession, a fluid core can only weakly mimic this motion.

For a precision of LLR ranging at cm-level, effects of general relativity are seen in the data. Parameters of relativity theory such as the famous post-Newtonian (PPN) β and γ parameters and the geodetic precession can be estimated (Williams et al. 1996 and 2004). For variation of the gravitational constant to be discernible, the precision on this determination must be better than at the present level of precision of the measurements, i.e. at the $1/c^2$ order of the gravitational gradient accelerations of Earth and Moon due to the Sun.

4 Direction of Future Clocks for Planetary Science

4.1 Advancements in Clocks Need to be Made for Space Science Instruments

For a clock to be onboard a spacecraft and to be beneficial for space science, its size and power-usage must be reduced as well as able to support launch, impacts, radiation, and large differences in temperature.

4.2 Limiting Factors of Advanced Clocks in Space Science

The limitation factor for the data using new clocks are related to other perturbations that are not yet sufficiently modeled and that contaminate the measurements, such as the plasma and atmosphere effects or the spacecraft motion. New methods have been discussed above that enable to circumvent these errors, either by using interferometric methods or using combination of different frequencies.

4.3 Known/Key Behavior for Clocks that Are Critical for Space Science

For future of clocks for space science, understanding the effect of thermal variation would be the enabling technology. Additionally, in any laser application, beam stability is primarily affected by point-to-point relative motion changes. This motion can be caused by many different sources of vibration and even acoustic sources. Analyses of the measured vibration-transfer functions can show resonance peaks generated by local vibration modes. The vibration modes may thus also be a challenging problem. They can re-introduce frequency noise with deleterious effects on the laser coherence that had been so carefully realized.

Concerning the stability, both the short-term and long-term stabilities are needed. For instance, the Allan variance of 10^{-13} at short term (@ 60 seconds) is necessary for the Doppler, and obtaining this with a miniaturized clock would already help the one-way link, assuming thermal and vibrational challenges are addressed. Long-term stability is necessary as well in order to be able to interpret long-term variations in the Doppler and ranging related to geophysics.

5 Conclusion

The planetary radioscience have addressed numerous important scientific objectives, such as the study of the gravity field for orbiting spacecraft or the demonstration of the presence of a liquid core inside a planet inferred from its rotation for a lander. Except for the atmospheric occultation science, the classical two-way radio links in X-band or Ka-band (or both) have shown to be sufficient as it relies on the frequency stability of the clocks of stations on Earth.

Currently, the precision of very good ultra-stable oscillator is at the level of 10^{-13}. Using such clocks in space would enable one-way links at the same precision as two-way links and reduces the geometric constraints of the observations, as well as improvement in the data noise since only one leg of radio path would be affected by the media error. Increasing the precision of the clock (better than 10^{-13}–10^{-14} stability) would immediately map into the precision of the measurements. However, at this level of precision, other error sources must be reduced as well in order to take the full benefit of clock's intrinsic accuracy. In orbiting or navigating spacecraft, non-gravitational accelerations acting on the spacecraft and contaminating the determination of the gravity field can be further estimated/corrected from precise accelerometer measurements. Additionally, the Earth media effects and plasma perturbations on the signal must be reduced to the level comparable to the accuracy of the new clock system. One way for calibrating radio links from media effects (and plasma effects if radio signal) is to consider interferometric methods like SBI or the new JPL laser facility. In that case, the large differential phase delay from the Earth's troposphere (and ionosphere and plasma for radio link) is a common mode bias, due to the small view angle of the transponders/laser beams as seen from the ground antenna ($\sim 2 \times 10^{-5}$ rad). It is therefore completely cancelled out. The interferometer setup does not allow the same cancellation for the other intervening media, namely the target planet or moon atmosphere and ionosphere if they exist. Another approach for media calibration is to use multi-frequency radio links, which can be used to almost completely remove the media errors.

Acknowledgements Part of the research was carried out at the Jet Propulsion Laboratory, California Institute of Technology, under a contract with the National Aeronautics and Space Administration.

References

J.B. Abshire, X. Sun, G. Neumann, J. McGarry, T. Zagwodzki, P. Jester, H. Riris, M. Zuber, D.E. Smith, Laser pulses from Earth detected at Mars, in *Conference on Lasers and Electro-Optics/Quantum Electronics and Laser Science Conference and Photonic Applications Systems Technologies, Technical Digest (CD)* (Optical Society of America, Washington, 2006), paper CThT6

S.W. Asmar, J.W. Armstrong, L. Iess, P. Tortora, Spacecraft Doppler tracking: noise budget and achievable accuracy in precision radio science observations. Radio Sci. **40**, RS2001 (2005). doi:10.1029/2004RS003101

R.-M. Baland, G. Tobie, A. Lefevre, T. Van Hoolst, Titan's internal structure inferred from its gravity field, shape, and rotation state. Icarus **237**, 29–41 (2014). doi:10.1016/j.icarus.2014.04.007

S. Bauer, H. Hussmann, J. Oberst, D. Dirkx, D. Mao, G.A. Neumann, E. Mazarico, M.H. Torrence, J.F. McGarry, D.E. Smith, M.T. Zuber, Demonstration of orbit determination for the Lunar Reconnaissance Orbiter using one-way laser ranging data. Planet. Space Sci. **129**, 32–46 (2016). doi:10.1016/j.pss.2016.06.005

S. Bauer, H. Hussmann, J. Oberst, D. Dirkx, D. Mao, G.A. Neumann, E. Mazarico, M.H. Torrence, J.F. McGarry, D.E. Smith, M.T. Zuber, Analysis of one-way laser ranging data to LRO, time transfer and clock characterization. Icarus **283**, 38–54 (2017). doi:10.1016/j.icarus.2016.09.026

P. Bender, Proposed microwave transponders for early lunar robotic landers. Adv. Space Res. **14**(6), 233–242 (1994). doi:10.1016/0273-1177(94)90033-7

D.M. Boroson, B.S. Robinson, D.V. Murphy et al., Overview and results of the lunar laser communication demonstration. Proc. SPIE **8971**, 1–11 (2014)

D. Buccino, J.A. Seubert, S.W. Asmar, R.S. Park, Optical ranging measurement with a lunar orbiter: limitations and potential. J. Spacecr. Rockets **53**(3), 457–463 (2016). doi:10.2514/1.A33415

J.J. Degnan, Millimeter accuracy satellite laser ranging: a review, in *Contributions of Space Geodesy to Geodynamics: Technology*. Monograph Geodynamics Series (AGU, Washington, 1993), pp. 133–162

J.M. Degnan, Asynchronous laser transponders for precise interplanetary ranging and time transfer. J. Geodyn. **34**, 551–594 (2002)

V. Dehant, W. Folkner, E. Renotte, D. Orban, S. Asmar, G. Balmino, J.P. Barriot, J. Benoist, R. Biancale, J. Biele, F. Budnik, S. Burger, O. de Viron, B. Häusler, Ö. Karatekin, S. Le Maistre, P. Lognonné, M. Menvielle, M. Mitrovic, M. Pätzold, A. Rivoldini, P. Rosenblatt, G. Schubert, T. Spohn, P. Tortora, T. Van Hoolst, O. Witasse, M. Yseboodt, Lander radioscience for obtaining the rotation and orientation of Mars. Planet. Space Sci. **57**, 1050–1067 (2009). doi:10.1016/j.pss.2008.08.009

V. Dehant, S. Le Maistre, A. Rivoldini, M. Yseboodt, P. Rosenblatt, T. Van Hoolst, M. Mitrovic, Ö. Karatekin, J.C. Marty, A. Chicarro, Revealing Mars' deep interior: future geodesy missions using radio links between landers, orbiters, and the Earth. Planet. Space Sci. **57**, 1069–1081 (2010). doi:10.1016/j.pss.2010.03.014

V. Dehant, B. Banerdt, P. Lognonné, M. Grott, S. Asmar, J. Biele, D. Breuer, F. Forget, R. Jaumann, C. Johnson, M. Knapmeyer, M. Lefeuvre, D. Mimoun, A. Mocquet, P. Read, A. Rivoldini, O. Romberg, G. Schubert, S. Smrekar, T. Spohn, P. Tortora, S. Ulamec, S. Vennerstrøm, Future Mars geophysical observatories for understanding its internal structure, rotation, and evolution. Planet. Space Sci. **68**(1), 123–145 (2012a). doi:10.1016/j.pss.2011.10.016

V. Dehant, J. Oberst, R. Nadalini, U. Schreiber, N. Rambaux, Geodesy instrument package on the Moon for improving our knowledge of the Moon and the realization of reference frames. Planet. Space Sci. **68**(1), 94–104 (2012b). doi:10.1016/j.pss.2012.02.008

D. Dirkx, R. Noomen, P.N.A.M. Visser, S. Bauer, L.L.A. Vermeersen, Comparative analysis of one- and two-way planetary laser ranging concepts. Planet. Space Sci. **117**, 159–176 (2015)

D. Dirkx, Interplanetary laser ranging—analysis for implementation in planetary science mission. PhD thesis, Delft University of Technology (2015)

D. Dirkx, R. Noomen, P.N.A.M. Visser, L.I. Gurvits, L.L.A. Vermeersen, Space-time dynamics estimation from space mission tracking data. Astron. Astrophys. **587**, A156 (2016)

M. Efroimsky, Bodily tides near spin-orbit resonances. Celest. Mech. Dyn. Astron. **112**(3), 283–330 (2012). doi:10.1007/s10569-011-9397-4

R.C. Elphic, C. Russell, *The Lunar Atmosphere and Dust Environment Explorer Mission (LADEE)* (Springer, Cham, 2015)

M. Fermi, P. Bender, B. Bertotti, M. Chersich, M. Gregnanin, L. Iess, L. Simone, Investigation of the lunar interior with a microwave interferometer. 37th COSPAR Scientific Assembly, paper P162-TueWed B01-0062-08 (2008)

W.M. Folkner, C.F. Yoder, D.N. Yuan, E.M. Standish, R.A. Preston, Interior structure and seasonal mass redistribution of Mars from radio tracking of Mars Pathfinder. Science **278**(5344), 1749 (1997)

M. Gregnanin, B. Bertotti, M. Chersich, M. Fermi, L. Iess, L. Simone, P. Tortora, J.G. Williams, Same beam interferometry as a tool for the investigation of the lunar interior. Planet. Space Sci. **74**, 194–201 (2012). doi:10.1016/j.pss.2012.08.027

M. Gregnanin, M. Yseboodt, V. Dehant, L. Iess, T. Van Hoolst, Estimation of Mars geophysical information through Same Beam Interferometry, in *Proc. European Planetary Science Congress EPSC 2014*, Centro de Congressos do Estoril, Cascais, Portugal Austria, 7–12 September, 2014, vol. 9 (2014), EPSC2014-EPSC395

H. Hemmati, K.M. Birnbaum, W.H. Farr, S. Turyshev, A. Biswas, Combined laser communications and laser ranging transponder for Moon and Mars, in *Free-Space Laser Communication Technologies XXI*. SPIE Conference Series, vol. 7199 (2009), No. 71990N

L. Iess, M. Fermi, P. Bender, B. Bertotti, M. Gregnanin, L. Simon, A microwave interferometer for the investigation of the lunar interior. NASA NOI N9-ILN09-0016 for the International Lunar Network (2008)

L. Iess, M.D. Benedetto, N. James, M. Mercolino, L. Simone, P. Tortora, Astra: interdisciplinary study on enhancement of the end-to-end accuracy for spacecraft tracking techniques. Acta Astronaut. **94**(2), 699–707 (2014)

A.S. Konopliv, C.F. Yoder, Venusian k_2 tidal Love number from Magellan and PVO tracking data. Geophys. Res. Lett. **23**(14), 1857–1860 (1996). doi:10.1029/96GL01589

A.S. Konopliv, S.W. Asmar, W.M. Folkner, Ö. Karatekin, D.C. Nunes, S.E. Smrekar, C.F. Yoder, M.T. Zuber, Mars high resolution gravity fields from MRO, Mars seasonal gravity, and other dynamical parameters. Icarus **211**, 401–428 (2011). doi:10.1016/j.icarus.2010.10.004

P. Kuchynka, W.M. Folkner, A.S. Konopliv, R.S. Park, S. Le Maistre, V. Dehant, New constraints on Mars rotation determined from radiometric tracking of the Opportunity Mars Exploration Rover. Icarus **222**(1), 243–253 (2013). doi:10.1016/j.icarus.2012.11.003

P. Kuchynka, W.M. Folkner, A.S. Konopliv, R.S. Park, S. Le Maistre, V. Dehant, New constraints on Mars rotation determined from radiometric tracking of the Opportunity Mars Exploration Rover. Icarus **229**, 340–347 (2014). doi:10.1016/j.icarus.2013.11.015

S. Le Maistre, P. Rosenblatt, A. Rivoldini, V. Dehant, J.C. Marty, Ö. Karatekin, Lander Radio science experiment with a direct link between Mars and the Earth. Planet. Space Sci. **68**(1), 105–122 (2012). doi:10.1016/j.pss.2011.12.020

D. Mao, J. McGarry, M. Torrence, G. Neumann, E. Mazarico, M. Barker, X. Sun, D. Rowlands, J. Golder, T. Zagwodzki, J. Cavanaugh, M. Zuber, D. Smith, Laser ranging experiment on Lunar Reconnaissance Orbiter: timing determination and orbit constraints, in *Proceedings of ILRS Workshop*, Bad Kotzing, Germany, May 2012 (2012). http://cddis.gsfc.nasa.gov/lw17/docs/papers/session13/02-Mao_LRO-LR_Kotzting2011_paper_final.pdf

D. Mao, J.F. McGarry, E. Mazarico, G.A. Neumann, X. Sun, M.H. Torrence, T.W. Zagwodzki, D.D. Rowlands, E.D. Hoffman, J.E. Horvath, J.E. Golder, M.K. Barker, D.E. Smith, M.T. Zuber, The laser ranging experiment of the Lunar Reconnaissance Orbiter: five years of operations and data analysis. Icarus **283**, 55–69 (2017). doi:10.1016/j.icarus.2016.07.003

E. Mazarico, F.G. Lemoine, S. Goossens, T.J. Sabaka, J.B. Nicholas, D.D. Rowlands, G.A. Neumann, M.H. Torrence, D.E. Smith, M.T. Zuber, Improved precision orbit determination of lunar orbiters from the GRAIL-derived lunar gravity models. Adv. Astronaut. Sci. **148**, 1125–1141 (2013)

G. Mitri, R. Meriggiola, A. Hayes, A. Lefevre, G. Tobie, A. Genova, J.I. Lunine, H. Zebkerg, Shape, topography, gravity anomalies and tidal deformation of Titan. Icarus **236**, 169–177 (2014). doi:10.1016/j.icarus.2014.03.018

W.B. Moore, G. Schubert, The tidal response of Ganymede and Callisto with and without liquid water oceans. Icarus **166**, 223–226 (2003). doi:10.1016/j.icarus.2003.07.001

T.D. Moyer, *Formulation for Observed and Computed Values of Deep Space Network Data Types for Navigation*. Monograph 2 Deep Space Communications and Navigation Series (2000). 549 pp.

T.W. Murphy, E.G. Adelberger, J.B.R. Battat, L.N. Carey, C.D. Hoyle, P. Leblanc, E.L. Michelsen, K. Nordtvedt, A.E. Orin, J.D. Strasburg, C.W. Stubbs, H.E. Swanson, E. Williams, The Apache Point Observatory lunar laser-ranging operation: instrument description and first detections. Publ. Astron. Soc. Pac. **120**, 20 (2008). doi:10.1086/526428

F. Nimmo, U.H. Faul, E.J. Garnero, Dissipation at tidal and seismic frequencies in a melt-free Moon. J. Geophys. Res. **117**(E9), E09005 (2012). doi:10.1029/2012JE004160

J. Oberst, V. Lainey, C. Le Poncin-Latte, V. Dehant, P. Rosenblatt, S. Ulamec, J. Biele, J. Spurmann, R. Kahle, V. Klein, U. Schreiber, A. Schlicht, N. Rambaux, P. Laurent, B. Noyelles, B. Foulon, A. Zakharov, L. Gurvits, D. Uchaev, S. Murchie, C. Reed, S.G. Turyshev, J. Gil, M. Graziano, K. Willner, K. Wickhusen, A. Pasewaldt, M. Wahlisch, H. Hussmann, GETEMME—a mission to explore the Martian satellites and the fundamentals of solar system physics. Exp. Astron. **34**, 243–271 (2012)

R.S. Park, A.S. Konopliv, S.W. Asmar, B.G. Bills, R.W. Gaskell, C.A. Raymond, D.E. Smith, M.T. Zuber, Gravity field expansion in ellipsoidal harmonic and polyhedral internal representations applied to Vesta. Icarus **240**, 118–132 (2013)

R.S. Park, A.S. Konopliv, B.G. Bills, N. Rambaux, J.C. Castillo-Rogez, C.A. Raymond, A.T. Vaughan, A.I. Ermakov, M.T. Zuber, R.R. Fu, M.J. Toplis, C.T. Russell, A. Nathues, F. Preusker, A partially differentiated interior for Ceres deduced from its gravity field and shape. Nature **537**, 515–517 (2016)

N. Rambaux, J.G. Williams, The Moon's physical librations and determination of their free modes. Celest. Mech. Dyn. Astron. **109**(1), 85–100 (2011). doi:10.1007/s10569-010-9314-2

D.E. Smith, M.T. Zuber, X. Sun, G.A. Neumann, J.F. McGarry, T.W. Zagwodzki, Two-way laser link over interplanetary distance. Science **311**, 53 (2006)

S.G. Turyshev, W. Farr, W.M. Folkner, A.R. Girerd, H. Hemmati, T.W. Murphy, J.G. Williams, J.J. Degnan, Advancing tests of relativistic gravity via laser ranging to Phobos. Exp. Astron. **28**(2), 209–249 (2010)

S.G. Turyshev, J.G. Williams, W.M. Folkner, G.M. Gutt, R.T. Baran, R.C. Hein, R.P. Somawardhana, J.A. Lipa, S. Wang, Corner-cube retro-reflector instrument for advanced lunar laser ranging. Exp. Astron. **36**(1–2), 105–135 (2013). doi:10.1007/s10686-012-9324-z

J.M. Wahr, M.T. Zuber, D.E. Smith, J.I. Lunine, Tides on Europa, and the thickness of Europa's icy shell. J. Geophys. Res. **111**(E12), E12005 (2006). doi:10.1029/2006JE002729

J.G. Williams, X.X. Newhall, J.O. Dickey, Relativity parameters determined from lunar laser ranging. Phys. Rev. D **53**, 6730–6739 (1996)

J.G. Williams, S.G. Turyshev, T.W. Murphy, Improving LLR tests of gravitational theory. Int. J. Mod. Phys. D **13**(3), 567–582 (2004). doi:10.1142/S0218271804004682

J.G. Williams, S.G. Turyshev, D.H. Boggs, Lunar laser ranging tests of the equivalence principle with the Earth and Moon. Int. J. Mod. Phys. D **18**(07), 1129–1175 (2009). doi:10.1142/S021827180901500X

J.G. Williams, A.S. Konopliv, D.H. Boggs, R.S. Park, D.-N. Yuan, F.G. Lemoine, S. Goossens, E. Mazarico, F. Nimmo, R.C. Weber, S.W. Asmar, H.J. Melosh, G.A. Neumann, R.J. Phillips, D.E. Smith, S.C. Solomon, M.M. Watkins, M.A. Wieczorek, J.C. Andrews-Hanna, J.W. Head, W.S. Kiefer, I. Matsuyama, P.J. McGovern, G.J. Taylor, M.T. Zuber, Lunar interior properties from the GRAIL mission. J. Geophys. Res. **119**(7), 1546–1578 (2014). doi:10.1002/2013JE004559

J.G. Williams, D.H. Boggs, Tides on the Moon: theory and determination of dissipation. J. Geophys. Res. **120**(4), 689–724 (2015). doi:10.1002/2014JE004755

X. Wu, Y.E. Bar-Sever, W.M. Folkner, J.G. Williams, J.F. Zumberge, Probing Europa's hidden ocean from tidal effects on orbital dynamics. Geophys. Res. Lett. **28**(11), 2245–2248 (2001). doi:10.1029/2000GL012814

S. Zhong, C. Qin, A. Geruo, J.M. Wahr, Can tidal tomography be used to unravel the long-wavelength structure of the lunar interior? Geophys. Res. Lett. **39**(15), L15201 (2012). doi:10.1029/2012GL052362

M.T. Zuber, D.E. Smith, R.S. Zellar, G.A. Neumann, X. Sun, R.B. Katz, I. Kleyner, A. Matuszeski, J.F. McGarry, M.N. Ott, L.A. Ramos-Izquierdo, D.D. Rowlands, M.H. Torrence, T.W. Zagwodzki, The lunar reconnaissance orbiter laser ranging investigation. Space Sci. Rev. **150**(1–4), 63–80 (2010)

The manufacturer's authorised representative in the EU is Springer Nature Customer Service Centre GmbH, Europaplatz 3, 69115 Heidelberg, Germany. If you have any concerns regarding our products, please contact ProductSafety@springernature.com

Printed and bound by CPI Group (UK) Ltd, Croydon, CR0 4YY
29/07/2026
02180519-0008